Linear and Non-linear Continuum Solid Mechanics

Santiago Hernández
University of A Coruña, Spain

&

Arturo N. Fontán
University of A Coruña, Spain

WIT*PRESS* Southampton, Boston

Santiago Hernández
University of A Coruña, Spain

&

Arturo N. Fontán
University of A Coruña, Spain

Published by

WIT Press
Ashurst Lodge, Ashurst, Southampton, SO40 7AA, UK
Tel: 44 (0) 238 029 3223; Fax: 44 (0) 238 029 2853
E-Mail: witpress@witpress.com
http://www.witpress.com

For USA, Canada and Mexico

Computational Mechanics International Inc
25 Bridge Street, Billerica, MA 01821, USA
Tel: 978 667 5841; Fax: 978 667 7582
E-Mail: infousa@witpress.com
http://www.witpress.com

British Library Cataloguing-in-Publication Data
A Catalogue record for this book is available
from the British Library

ISBN: 978-1-78466-271-4
eISBN: 978-1-78466-272-1

Library of Congress Catalog Card Number: 2017958182

About the Authors

Joe Pulizzi, author, speaker, and evangelist for content marketing, is dedicated to helping companies grow their profits by creating better content. One of the founders of the content marketing movement, Joe launched Junta42.com in 2007 as a true online resource for those interested in content marketing and custom publishing. Junta42's Match service is the industry's only online marketplace dedicated to custom publishing and content marketing. For his efforts, Joe was voted Custom Media Innovator of the Year in 2008 by American Business Media. Formerly vice president of custom media for Penton Media, Inc., Joe serves on the board of the Custom Publishing Council and was two-time Custom Media Chair for American Business Media.

Newt Barrett is president of Content Marketing Strategies, a division of Voyager Media, Inc. Content Marketing Strategies helps small and medium-sized companies grow sales by delivering compelling and relevant content that targets their customers precisely. Newt is also the blogger-in-chief at ContentMarketingToday.com, an educational Web site dedicated to helping marketers find new ways to reach their customers in print, online, and in person. Newt is a successful publishing executive with more than 25 years of experience, having launched publications for CMP, Ziff-Davis, Penton Media, Inc., and Southwest Florida Business.

For more information, please visit www.getcontentgetcustomers.com.
More of Joe at blog.junta42.com.
More of Newt at www.ContentMarketing.net.

Dedicated to our families:

Pilar, Guillermo, Carlos and Dani

Ro, Lois and Carme

CONTENTS

Preface xi

1 Introduction 1
1.1 Continuum Mechanics 1
1.2 The Continuum Deformable Solid 2

2 Relationships Between Displacements and Strains in Deformable Bodies. Kinematic Equations 7
2.1 Displacements Field in Deformable Bodies 7
2.2 Strain Tensors 9
2.2.1 Tensors in large strain theory 9
2.2.2 Tensors in small strain theory: Cauchy strain tensor . . 12
2.2.3 Another definition of strain: Cauchy–Green strain tensor 15
2.3 Principal Directions of Strain and Maximum Angular Strain . 16
2.4 Ellipsoid of Strains 22
2.5 Indicatrix Quadric of Strains 23
2.6 Directrix Quadric of Strains 27
2.7 Mohr's Circles of Strain 28
2.8 Compatibility Equations in Theory of Small Strains 31

3 Equilibrium Equations in Deformable Bodies 37
3.1 Stress Tensor at a Point 37
3.2 Internal Equilibrium Equations 39
3.3 Equations of Equilibrium on the Boundary 43
3.4 Principal Stresses and Principal Stress Directions 44
3.5 Maximum Shear Stresses 46
3.6 Quadrics Associated with the Stress Tensor 48
3.6.1 Ellipsoid of stress 48
3.6.2 Indicatrix quadric of stresses 49
3.6.3 Directrix quadric of stresses 52
3.7 Mohr's Circles of Stress 53

3.8 Spherical and Deviatoric Tensors 55

4 Relationship Between Stresses and Strains in Deformable Bodies. Constitutive Equations 63
4.1 Behaviour of Continuous Bodies. Constitutive Equations . . . 63
4.2 Linear Elastic Materials. Generalized Hooke's Law 65
4.3 Strain Energy 73

5 Plane Linear Elasticity. Plane Strain and Plane Stress 77
5.1 Plane Strain Field 77
5.2 Plane Stress Field 79
5.3 Principal Directions in Plane Elasticity 81
5.4 Mohr's Circle in Plane Elasticity 82
5.5 Stress Function in Plane Elasticity. Airy Function 85
5.5.1 Plane strain field 85
5.5.2 Plane stress field 86
5.6 Representative Points and Lines in Plane Elasticity 87
5.7 Plane Elasticity in Polar Coordinates 89
5.7.1 Equilibrium equations and kinematic relationships . . 89
5.7.2 Plane stress and plane strain fields 93
5.7.3 Airy function in polar coordinates 95
5.8 Examples of Elastic Problems in Polar Coordinates 96
5.8.1 Circular tube solicited by radial pressure 96
5.8.2 Wedge and semi-infinite continuum with a concentrated vertical load 100
5.8.3 Semi-infinite domain with a distributed load 104
5.8.4 Thin plate with hole 106

6 Hyperelasticity 117
6.1 Hyperelastic Materials 117
6.2 Models of Incompressible Hyperelastic Materials 118
6.2.1 Mooney–Rivlin model 119
6.2.2 Neo-Hooke model 119
6.2.3 Yeoh model 119
6.2.4 Ogden model 120
6.3 Strain Energy in Hyperelastic Materials 120
6.4 Obtaining the Stress Tensor in Hyperelastic Materials 121
6.5 Stress Tensors in Incompressible Hyperelastic Materials. Uniaxial Forces 122
6.5.1 Stress in the Mooney–Rivlin model 124
6.5.2 Stress in the Neo-Hooke model 124
6.5.3 Stress in the Yeoh model 124

6.5.4 Stress in the Ogden model 124
6.6 Compressible Hyperelastic Materials 125
6.6.1 Mooney–Rivlin model 126
6.6.2 Neo-Hooke model 127
6.6.3 Ogden model 127
6.6.4 Yeoh model 127

7 Plasticity **131**
7.1 Introduction 131
7.2 Plastification Criteria in Metallic Materials 131
7.2.1 Beltrami–Haigh criterion 131
7.2.2 Von Mises–Hencky criterion 133
7.2.3 Tresca criterion 136
7.3 Plastification Criteria in Soil Mechanics 138
7.3.1 Mohr–Coulomb criterion 138
7.3.2 Drucker–Prager criterion 140
7.3.3 Cambridge model. Cam Clay and modified Cam Clay 142
7.4 Elastoplastic Behaviour of Materials 145
7.4.1 Elastoplastic behaviour without strain hardening 146
7.4.2 Elastoplastic behaviour with strain hardening 148

8 Linear Viscoelasticity **155**
8.1 Introduction 155
8.2 Models of Viscoelastic Behaviour 155
8.2.1 Two-parameter models. Maxwell and Voigt models . . 156
8.2.2 Three-parameter models. Maxwell's and Voigt's standard linear models 160
8.2.3 Generalized standard linear models. Maxwell and Voigt models 167

9 Linear Elasto-viscoplasticity **177**
9.1 Introduction 177
9.2 Model with Two Parameters 177
9.3 Models with Three Parameters. Bingham Models 179
9.4 Bingham's Four-Parameter Model 183
9.5 Non-linear Viscosity Models. Norton's Model 188

Bibliography **191**

PREFACE

Continuum mechanics studies of deformable solids and fluids have always been part of the curriculum of civil, mechanical and aeronautical engineering degrees, amongst others. More recently, they have been included in other programmes, such as materials and biomechanics engineering. This book takes into consideration topics related to deformable solids, that is to say, those that undergo changes in geometry when subjected to external loads or other types of deformation. The objective of this volume is to present material that can be used for teaching continuum mechanics to students of the disciplines mentioned above. In order to understand the contents, the reader only needs to know linear algebra and differential calculus.

In the past, books such as this only described linear elasticity and plasticity theories of a homogeneous and isotropic continuum. Linear elasticity can only represent in an approximate manner the behaviour of the materials most commonly used in engineering. Its advantage is not only its simplicity but also that, over time, it has given rise to a wide range of analytical solutions. This was an important consideration in the period prior to the development and availability of powerful computers. Plasticity is able to represent the ductile behaviour of materials such as steel or different types of soil, for instance.

This book deals with advanced linear elasticity and plasticity approaches as well as the study of the behaviour of more complex types of materials, such as those of more recent manufacture, namely, composites and others whose material characterisation has been possible only recently. It describes how linear elastic behaviour extends to anisotropic materials in general and how deformations can result in small or large strain components. Plastic behaviour is also extended to include the strain hardening of materials.

Among others, new topics incorporated in this volume are studies of hyperelastic materials, which can represent elastomeric materials and some types of biological materials. A section of the volume deals with viscoelastic materials, i.e. those which deform when subjected to long-term loads. The cases of viscoplastic as well as elasto-viscoplastic behaviour describe other types of materials well, including those present in many geotechnical sites.

Examples have been included throughout the text and, at the end of each chapter, exercises are presented that can be used by students to check their comprehension of the theoretical information presented. The authors hope that the contents of the book will be of interest to researchers and students, and they will be grateful to receive any comments that readers can provide to improve the text in future editions. They want to thank the late Prof. C.A. Brebbia for his careful review of the manuscript. We feel honoured for that.

Finally, the authors want to express their appreciation to Xián Meirás for his help in the editing of the book and to Natalia Calvo for her translation of the original Spanish text into English and to WIT Press for accepting the manuscript for publication.

The Authors
La Coruña, 2021

CHAPTER 1

INTRODUCTION

1.1 CONTINUUM MECHANICS

Mechanics is the part of physics that studies the motion and equilibrium of bodies, as well as the forces that produce them, and within this science, continuum mechanics is the branch dedicated to material bodies, as opposed to the mechanics of the material point. Continuum mechanics proposes models for deformable solids and fluids.

Continuous material is understood to be an idealized material whose properties can be defined mathematically as continuous functions in space and time. The matter is composed of discrete particles linked together by more or less intense forces that determine its solid, liquid or gaseous condition. However, an atomic size analysis lacks interest from an engineering point of view because what is important is that the results obtained, by applying the concept of continuum, are consistent with the observed macroscopic behaviour.

If a body acts as a continuum, then regardless of how small the portion under consideration is, the properties of the material studied will be the same as the original material. The continuum assumption can reduce an arbitrary volume to a material point, thus allowing quantities that are of interest to be defined, for example density, stress and deformation.

The physical properties addressed by continuum mechanics are independent of the coordinate system with which they are observed; thus, they can be represented by tensors, mathematical concepts that have the attribute of being independent of the reference system.

Continuum mechanics determines the internal behaviour of a body in response to the external forces acting on it and provides the basis for a variety of scientific fields, such as mechanics of materials, strength of materials, hydraulics, material science, rock mechanics, hydrodynamics, aeroelasticity and biomechanics. In disciplines such as rock mechanics, the hypothesis of continuous mechanics is not completely valid, and it is therefore necessary to complement continuum mechanics with specific theories.

The behaviour of a continuum model can be described by applying the fundamental laws of physics, such as the conservation of mass, the conservation of momentum or the conservation of energy, and the principles of behaviour of each idealized material, which are called constitutive equations and which take into account their particular characteristics.

As regards methods of study, a continuum can be analysed by theoretical formulations, trial tests in a laboratory using scale models and/or by computational methods, for instance, numerical models. In this last case, computational mechanics and the huge advances in digital computers have enabled problems considered unsolvable until only a few decades ago to be solved.

Continuums can be classified according to their mechanical behaviour into deformable solids and fluids, whose fundamental difference is that, in the former, the strain at a precise point and instant depend on the deformation of that point, that is to say, the difference between the initial and final setup, while in fluids, the strain at a given point and instant depend on the pressure and deformation speed, but not on the deformation itself.

Rheology, which marks the fuzzy border between deformable solid mechanics and fluid mechanics, is the science that studies materials that initially behave like solids, but with time flow like fluids.

The parameters that characterize the fluid material are viscosity, which is the internal force that opposes its displacement, and density. Fluids can be classified according to their viscosity as non-viscous fluids, perfect Newtonian fluids and non-Newtonian fluids, and according to their density as compressible and incompressible.

The fundamental concepts of continuum mechanics exclusively applied to deformable solids will be introduced in this book.

1.2 THE CONTINUUM DEFORMABLE SOLID

The mechanics of deformable solids is the branch of physics that addresses continuums that have a definite form, as opposed to fluids, whose form is determined entirely by the recipient or the set of restrictions on its surface and which require different studies.

While studying continuums it is always necessary to consider three types of equations: kinematic equations, equilibrium equations and the material constitutive law.

The formulation of the kinematic equations begins with the position changes of the continuum points, the motion. With these variations that they experience, it is worth defining the deformations, which are a fundamental aspect in engineering, to determine the condition of a material therefore, the ultimate goal of these equations is to relate the displacements and the deformations of the material points. There are various approaches to defining the concept of deformation, which will be noted throughout the text.

It is important to be aware that the degrees of motion experienced by a continuum due to the action of a load can be of different magnitudes: in some cases it may be small, and consequently its geometry is very slightly modified; however, situations exist in which the opposite happens. In this case, this leads to deciding the reference system of the continuum points and offers two possible approaches: Lagrange's, which uses the initial reference system, or Euler's, which uses the one that defines a continuum after the volume changes, and both are developed later. In small deformations, both approaches blend with the Cauchy model. Given that the deformations of a continuum have a tensor character, all the above can be defined by a set of tensors, Lagrange–Green, Euler–Almansi or Cauchy–Green, which describe the deformation of the continuum in different ways.

The equilibrium relationships are of two types, those equations that associate the stresses inside the material with the mass loads that it is subjected to, and which are referred to as internal equilibrium equations; and those that relate the stress on the surface of the body's boundary with the distributed loads that exist there, and which are called boundary equilibrium equations. As in the deformations, it is convenient to formulate different stress tensors – Cauchy, Kirchhoff, Piola–Kirchhoff – that facilitate treating the stress field of a continuum. Some of those possible fields are more relevant than others, especially those called spherical and deviatoric, that are used in the criteria of material plasticity, and hence will be described in detail.

The remaining group of equations, also called the constitutive laws, are relationships between two different magnitudes of the continuum, stresses and deformations, and therefore tend to be described as mixed relationships. The constitutive equations do not derive from the fundamental laws of physics, but are specific to each material and are formulated based on the observation of their behaviour while being subjected to a set of exterior actions.

The materials used in engineering are very varied: steel, concrete, soils and rocks, wood, aluminium, elastomeric, carbon fibre and fibreglass, among others. Their behaviours are very different, and the corresponding constitutive equations must be defined for each of them.

Historically, materials were studied as homogeneous and isotropic, which is to say that it was assumed that their compositions were identical in any of the volume points and that their mechanical properties were the same in any direction. However, in some of them it is obvious that their mechanical response varies according to their orientation, and therefore require more refined formulations. Thus, orthotropic formulations that define the constitutive equations of the material in a system of three-dimensional and orthogonal axes are usually considered. This contributes to the study being performed more according to reality.

This approach is very useful in materials that are composed of fibres that go in certain directions, as in the case of wood, or in modern materials made of glass or carbon fibres. It is also suitable for materials that are homogeneous but have holes in some direction; this circumstance varies the strength.

A continuum that is considered elastic and linear can be studied using linear elasticity theory. In this case a rule of proportionality exists between the stresses and deformations in the material, and the material's response to several loading conditions can be considered as the combined response to each of them, which is known as the principle of superposition. This formulation has a deep-rooted tradition in analytical solutions to problems of stress fields and is the foundation of the theorems that have been used in the elastic and linear analysis of all types of structures.

Among the possible conditions of linear elasticity, the cases of plane stress and plane deformation are the best known because they represent many situations that occur in real life with sufficient approximation. In both cases, the components of the stress fields can be obtained from second-order partial differential equations using ad-hoc expressions, known as stress functions and often as Airy functions, in honour of their discoverer. From the stress tensor, the deformation tensor and the displacement expressions of the continuum can be obtained at any point.

To carry out this study in an analytical way, it is convenient to use an adequate coordinate system, and this depends on the geometry of the domain that is being considered. For this reason, on some occasions Cartesian coordinates are used, and on others the polar coordinate system is more convenient. The latter can be more appropriate in situations in which there is symmetry between a perpendicular axis and the studied domain and in those in which semi-infinite regions are analysed. Some of these circumstances occur in problems related to soil mechanics; thus, this text will offer examples of situations of this kind.

On the other hand, there are other elastic materials in which such proportionality does not occur. These are known as hyperelastic materials, and include organic materials such as rubbers or biological materials such as the tissues of living beings. The latter are becoming increasingly important due to the development of the field of biomedical engineering, which studies the mechanical behaviour of these materials. Among the hyperelastic materials, it is important to distinguish between the incompressible and those experiencing a change in volume as they are deformed, i.e. the compressible.

As will be explained later in this book, previous hyperelasticity models assumed the condition of volume conservation, that is to say, the materials were considered incompressible because the elastomeric materials, for instance, responded well to this hypothesis, and formulations such as the neo-Hooke, Yeoh, Ogden or Mooney–Rivlin provided adequate results. However, this does not occur in the materials making up the tissues of living beings; therefore, those models have evolved into updated versions to adapt themselves to the general behaviour of hyperelastic compressible materials.

It is very common for a material's elastic phase to finish at some point and after that change its behaviour, which generally leads to deformation growth without requiring a stress increase, giving rise to the phenomenon known as plasticity. What triggers this phenomenon – that is to say, the stress field that ends the elastic phase – is very diverse and differs between different materials. This also occurs in geological materials, and for this reason soil mechanics has always taken into consideration the plasticity of soil or rock.

Given its phenomenological character, this behaviour was studied using experimental measurements in which the material was subjected to a specific set of loading conditions, and also theoretical approaches that tried to reproduce the results from the tests in a reliable manner. The latter had the advantage of being more general because they were capable of identifying the start of the plasticity in any load. The criteria of Beltrami–Haigh, Von Mises–Hencky or Treska, and those of Mohr–Coulomb, Druker–Prager or those of Cam Clay in the field of soil mechanics, are the best known models of metallic materials.

In cyclical loads, the plastic phase stops if the value of the load diminishes, and the material may return to a new elastic phase, giving rise to what is called elastoplastic behaviour. Thus, the material can maintain its previous properties or, on the contrary, may be able to undergo higher stresses. This behaviour is called hardening by deformation.

There are other material responses that are of interest, for example what occurs when the deformation of a continuum increases over time due to a constant loading condition, known as the creep phenomenon, or what occurs when a constant deformation condition can take place with a decreasing value of stress, known as the relaxation phenomenon. The first commonly occurs in concrete and the second in steel when it is subjected to an initial strain, which is normal in prestressed concrete structures or the cables of cable-stayed bridges.

Both belong to the field of viscoelasticity and combine the behaviour of elastic solids with fluids. To study the types of time-dependent responses of materials, there are mathematical models composed of the simple models of Maxwell or Voigt, which are combined in an order of increasing complexity to accurately reproduce the behaviour of those materials in real life.

The final constitutive model that is presented in this text is that of elasto-viscoplastic materials. These have an initial elastic phase, and once it is overcome a permanent and increased deformation is produced. Fresh concrete or clay soils are examples of these materials. This text exclusively studies one example of this type of material in which the elastic and viscous components are linear. The Bingham model is among those most often used to analyse this type of deformability, and several of its versions, using three and four parameters, are described in detail in the text.

CHAPTER 2

RELATIONSHIPS BETWEEN DISPLACEMENTS AND STRAINS IN DEFORMABLE BODIES. KINEMATIC EQUATIONS

2.1 DISPLACEMENTS FIELD IN DEFORMABLE BODIES

Suppose the deformable body shown in Figure 2.1.1 is loaded by a set of external actions. As a consequence, the body will be deformed and its points will have changed position. For example, the point $P(x, y, z)$, defined by the vector $\mathbf{x}$, will be at the position $P_1(r_1, r_2, r_3)$, defined by the vector $\mathbf{r}$. The relationships between the coordinates of both points are

$$r_1 = x + u \quad r_2 = y + v \quad r_3 = z + w \tag{2.1.1}$$

or in vector form

$$\mathbf{r} = \mathbf{x} + \mathbf{u} \tag{2.1.2}$$

with u, v, w being the variation of the coordinates of the point in each axis.

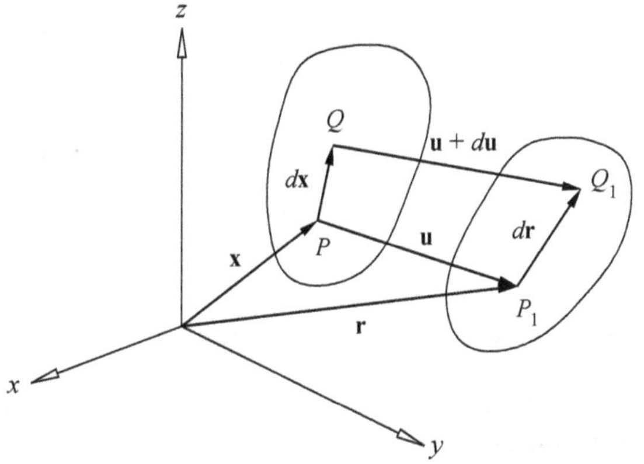

Figure 2.1.1: Displacements of a point of a deformable body

The coordinates $\mathbf{x}$ are usually called *material* or *Lagrangian coordinates* and attend to the initial positions of the points on the body. The coordinates $\mathbf{r}$ refer to the modified positions of the points on the body and are called *spatial* or *Eulerian coordinates.*

In the displacements field, the differential line element $d\mathbf{x}$ defined by the points PQ will become the differential line element $d\mathbf{r}$ defined by the points P_1Q_1, and the relationship will become

$$\mathbf{r} + d\mathbf{r} = \mathbf{x} + d\mathbf{x} + \mathbf{u} + d\mathbf{u} \tag{2.1.3}$$

and, recalling (2.1.2), will result in

$$d\mathbf{r} = d\mathbf{x} + d\mathbf{u} \tag{2.1.4}$$

From (2.1.1) the following matrix can be written

$$\begin{bmatrix} dr_1 \\ dr_2 \\ dr_3 \end{bmatrix} = \begin{bmatrix} 1 + \dfrac{\partial u}{\partial x} & \dfrac{\partial u}{\partial y} & \dfrac{\partial u}{\partial z} \\ \dfrac{\partial v}{\partial x} & 1 + \dfrac{\partial v}{\partial y} & \dfrac{\partial v}{\partial z} \\ \dfrac{\partial w}{\partial x} & \dfrac{\partial w}{\partial y} & 1 + \dfrac{\partial w}{\partial z} \end{bmatrix} \begin{bmatrix} dx \\ dy \\ dz \end{bmatrix} \tag{2.1.5}$$

or also

$$d\mathbf{r} = \left(\mathbf{I} + \frac{\partial \mathbf{u}}{\partial \mathbf{x}}\right) d\mathbf{x} = (\mathbf{I} + \mathbf{J}_d)\, d\mathbf{x} \tag{2.1.6}$$

where the matrix $\mathbf{J}_d$ contains the first derivatives of the displacements of point P. For this reason it is known as the *matrix of the displacement vector gradients.*

$$\mathbf{J}_d = \frac{\partial \mathbf{u}}{\partial \mathbf{x}} = \begin{bmatrix} \dfrac{\partial u}{\partial x} & \dfrac{\partial u}{\partial y} & \dfrac{\partial u}{\partial z} \\ \dfrac{\partial v}{\partial x} & \dfrac{\partial v}{\partial y} & \dfrac{\partial v}{\partial z} \\ \dfrac{\partial w}{\partial x} & \dfrac{\partial w}{\partial y} & \dfrac{\partial w}{\partial z} \end{bmatrix} = \begin{bmatrix} \mathbf{u}_x & \mathbf{u}_y & \mathbf{u}_z \end{bmatrix} \tag{2.1.7}$$

with $\mathbf{u}_x$, $\mathbf{u}_y$, $\mathbf{u}_z$ being column vectors that contain the first derivatives of the vector components of the displacements with respect to each of the coordinate axes.

Also, the vector P_1Q_1, defined as $d\mathbf{r}$, can be obtained according to the vector $d\mathbf{x}$ by using the expression

$$d\mathbf{r} = \frac{\partial \mathbf{r}}{\partial \mathbf{x}} d\mathbf{x} = \mathbf{J} d\mathbf{x} \tag{2.1.8}$$

where $\mathbf{J}$ is the Jacobian matrix in the reference systems $\mathbf{x}$ and $\mathbf{r}$ whose elements are

$$\mathbf{J} = \frac{\partial \mathbf{r}}{\partial \mathbf{x}} = \begin{bmatrix} \frac{\partial r_1}{\partial x} & \frac{\partial r_1}{\partial y} & \frac{\partial r_1}{\partial z} \\ \frac{\partial r_2}{\partial x} & \frac{\partial r_2}{\partial y} & \frac{\partial r_2}{\partial z} \\ \frac{\partial r_3}{\partial x} & \frac{\partial r_3}{\partial y} & \frac{\partial r_3}{\partial z} \end{bmatrix} = \begin{bmatrix} \mathbf{r}_x & \mathbf{r}_y & \mathbf{r}_z \end{bmatrix} \tag{2.1.9}$$

and its determinant J can be expressed as

$$J = \mathbf{r}_x \cdot \left(\mathbf{r}_y \times \mathbf{r}_z \right) \tag{2.1.10}$$

where $\mathbf{r}_x$, $\mathbf{r}_y$, $\mathbf{r}_z$ are column vectors that contain the first derivatives of the vector that defines the position of the point P_1 with respect to each coordinate axis. Therefore, the matrix $\mathbf{J}$ is also called the *matrix of the position vector gradients.* Comparing the expressions (2.1.6) and (2.1.9), one can verify that

$$\mathbf{J} = \mathbf{I} + \mathbf{J}_d \tag{2.1.11}$$

2.2 STRAIN TENSORS

2.2.1 Tensors in large strain theory

The strain of the points on the body can be studied from the variations of the differential line elements PQ and P_1Q_1, defined by the vectors $d\mathbf{x}$ and $d\mathbf{r}$ and whose lengths are dl and dl_r, respectively. They accomplish

$$dl^2 = d\mathbf{x}^T d\mathbf{x} = dx^2 + dy^2 + dz^2 \tag{2.2.1.1}$$

and

$$dl_r^2 = d\mathbf{r}^T d\mathbf{r} = dr_1^2 + dr_2^2 + dr_3^2 \tag{2.2.1.2}$$

recalling equation (2.1.8)

$$dl_r^2 = d\mathbf{x}^T \mathbf{J}^T \mathbf{J} d\mathbf{x} \tag{2.2.1.3}$$

the differences between both lengths are

$$dl_r^2 - dl^2 = d\mathbf{r}^T d\mathbf{r} - d\mathbf{x}^T d\mathbf{x} = d\mathbf{x}^T \left(\mathbf{J}^T \mathbf{J} - \mathbf{I} \right) d\mathbf{x} \tag{2.2.1.4}$$

which can be written as

$$dl_r^2 - dl^2 = 2d\mathbf{x}^T \mathbf{E}_l d\mathbf{x} \tag{2.2.1.5}$$

where $\mathbf{E}_l$ is a tensor. It is associated with the strain of the differential line element using the initial position of the point on the body and is called the

Green–Lagrange strain tensor. Recalling equation (2.1.9), one can write it as

$$\mathbf{E}_l = \frac{1}{2}\left(\mathbf{J}^T\mathbf{J} - \mathbf{I}\right) = \frac{1}{2}\begin{bmatrix} \mathbf{r}_x{}^T\mathbf{r}_x - 1 & \mathbf{r}_x{}^T\mathbf{r}_y & \mathbf{r}_x{}^T\mathbf{r}_z \\ \mathbf{r}_x{}^T\mathbf{r}_y & \mathbf{r}_y{}^T\mathbf{r}_y - 1 & \mathbf{r}_y{}^T\mathbf{r}_z \\ \mathbf{r}_x{}^T\mathbf{r}_z & \mathbf{r}_y{}^T\mathbf{r}_z & \mathbf{r}_z{}^T\mathbf{r}_z - 1 \end{bmatrix} \tag{2.2.1.6}$$

The difference between dl_r^2 and dl^2 can also be expressed according to $d\mathbf{r}$. Recalling equation (2.1.8)

$$d\mathbf{x} = \mathbf{J}^{-1}d\mathbf{r} \tag{2.2.1.7}$$

and

$$dl_r^2 - dl^2 = d\mathbf{r}^T d\mathbf{r} - d\mathbf{x}^T d\mathbf{x} = d\mathbf{r}^T\left(\mathbf{I} - \mathbf{J}^{-1^T}\mathbf{J}^{-1}\right)d\mathbf{r} \tag{2.2.1.8}$$

or also

$$dl_r^2 - dl^2 = 2d\mathbf{r}^T\mathbf{E}_e d\mathbf{r} \tag{2.2.1.9}$$

where

$$\mathbf{E}_e = \frac{1}{2}\left(\mathbf{I} - \mathbf{J}^{-1^T}\mathbf{J}^{-1}\right) \tag{2.2.1.10}$$

with $\mathbf{E}_e$ being a symmetric tensor, whose elements are dimensionless, and is associated with the strain of the differential line element, using the final positions of the points on the body. It is named the *Euler–Almansi strain tensor*.

The tensors $\mathbf{E}_l$ and $\mathbf{E}_e$ present the relationship between them

$$\mathbf{E}_l = \mathbf{J}^T\mathbf{E}_e\mathbf{J} \tag{2.2.1.11}$$

Returning to the expressions (2.2.1.1) and (2.2.1.2) and recalling (2.1.2), one has

$$dl_r^2 = dx^2 + 2dxdu + du^2 + dy^2 + 2dydv + dv^2 + dz^2 + 2dzdw + dw^2 \tag{2.2.1.12}$$

$$dl_r^2 - dl^2 = 2\left(dxdu + dydv + dzdw\right) + du^2 + dv^2 + dw^2 \tag{2.2.1.13}$$

the derivatives *du, dv, dw* are

$$du = \frac{\partial u}{\partial x}dx + \frac{\partial u}{\partial y}dy + \frac{\partial u}{\partial z}dz \tag{2.2.1.14a}$$

$$dv = \frac{\partial v}{\partial x}dx + \frac{\partial v}{\partial y}dy + \frac{\partial v}{\partial z}dz \tag{2.2.1.14b}$$

$$dw = \frac{\partial w}{\partial x}dx + \frac{\partial w}{\partial y}dy + \frac{\partial w}{\partial z}dz \tag{2.2.1.14c}$$

substituting in (2.2.1.13)

$$
\begin{aligned}
dl_r^2 - dl^2 = {} & 2\left\{\frac{\partial u}{\partial x} + \frac{1}{2}\left[\left(\frac{\partial u}{\partial x}\right)^2 + \left(\frac{\partial v}{\partial x}\right)^2 + \left(\frac{\partial w}{\partial x}\right)^2\right]\right\} dx^2 \\
& + 2\left\{\frac{\partial v}{\partial y} + \frac{1}{2}\left[\left(\frac{\partial u}{\partial y}\right)^2 + \left(\frac{\partial v}{\partial y}\right)^2 + \left(\frac{\partial w}{\partial y}\right)^2\right]\right\} dy^2 \\
& + 2\left\{\frac{\partial w}{\partial z} + \frac{1}{2}\left[\left(\frac{\partial u}{\partial z}\right)^2 + \left(\frac{\partial v}{\partial z}\right)^2 + \left(\frac{\partial w}{\partial z}\right)^2\right]\right\} dz^2 \\
& + 2\left(\frac{\partial v}{\partial x} + \frac{\partial u}{\partial y} + \frac{\partial u}{\partial x}\frac{\partial u}{\partial y} + \frac{\partial v}{\partial x}\frac{\partial v}{\partial y} + \frac{\partial w}{\partial x}\frac{\partial w}{\partial y}\right) dxdy \\
& + 2\left(\frac{\partial w}{\partial x} + \frac{\partial u}{\partial z} + \frac{\partial u}{\partial x}\frac{\partial u}{\partial z} + \frac{\partial v}{\partial x}\frac{\partial v}{\partial z} + \frac{\partial w}{\partial x}\frac{\partial w}{\partial z}\right) dxdz \\
& + 2\left(\frac{\partial w}{\partial y} + \frac{\partial v}{\partial z} + \frac{\partial u}{\partial y}\frac{\partial u}{\partial z} + \frac{\partial v}{\partial y}\frac{\partial v}{\partial z} + \frac{\partial w}{\partial y}\frac{\partial w}{\partial z}\right) dydz
\end{aligned}
\tag{2.2.1.15}
$$

This expression can be represented in matrix form as

$$
dl_r^2 - dl^2 = 2\begin{bmatrix} dx & dy & dz \end{bmatrix}\begin{bmatrix} e_{11} & e_{12} & e_{13} \\ e_{21} & e_{22} & e_{23} \\ e_{31} & e_{32} & e_{33} \end{bmatrix}\begin{bmatrix} dx \\ dy \\ dz \end{bmatrix} \tag{2.2.1.16a}
$$

or

$$
dl_r^2 - dl^2 = 2d\mathbf{x}^T \mathbf{E}_l d\mathbf{x} \tag{2.2.1.16b}
$$

Recalling (2.2.1.5), one sees that these elements are those of the Green–Lagrange tensor, being

$$
e_{11} = \frac{\partial u}{\partial x} + \frac{1}{2}\left[\left(\frac{\partial u}{\partial x}\right)^2 + \left(\frac{\partial v}{\partial x}\right)^2 + \left(\frac{\partial w}{\partial x}\right)^2\right] \tag{2.2.1.17a}
$$

$$
e_{22} = \frac{\partial v}{\partial y} + \frac{1}{2}\left[\left(\frac{\partial u}{\partial y}\right)^2 + \left(\frac{\partial v}{\partial y}\right)^2 + \left(\frac{\partial w}{\partial y}\right)^2\right] \tag{2.2.1.17b}
$$

$$e_{33} = \frac{\partial w}{\partial z} + \frac{1}{2}\left[\left(\frac{\partial u}{\partial z}\right)^2 + \left(\frac{\partial v}{\partial z}\right)^2 + \left(\frac{\partial w}{\partial z}\right)^2\right] \tag{2.2.1.17c}$$

$$e_{12} = e_{21} = \left(\frac{\partial u}{\partial y} + \frac{\partial v}{\partial x} + \frac{\partial u}{\partial x}\frac{\partial u}{\partial y} + \frac{\partial v}{\partial x}\frac{\partial v}{\partial y} + \frac{\partial w}{\partial x}\frac{\partial w}{\partial y}\right)\frac{1}{2} \tag{2.2.1.17d}$$

$$e_{13} = e_{31} = \left(\frac{\partial u}{\partial z} + \frac{\partial w}{\partial x} + \frac{\partial u}{\partial x}\frac{\partial u}{\partial z} + \frac{\partial v}{\partial x}\frac{\partial v}{\partial z} + \frac{\partial w}{\partial x}\frac{\partial w}{\partial z}\right)\frac{1}{2} \tag{2.2.1.17e}$$

$$e_{23} = e_{32} = \left(\frac{\partial v}{\partial z} + \frac{\partial w}{\partial y} + \frac{\partial u}{\partial y}\frac{\partial u}{\partial z} + \frac{\partial v}{\partial y}\frac{\partial v}{\partial z} + \frac{\partial w}{\partial y}\frac{\partial w}{\partial z}\right)\frac{1}{2} \tag{2.2.1.17f}$$

It is convenient to also know the relationship between the initial and final volumes, V and V_r, respectively. If one considers an elementary volume of dimensions dx, dy, dz, in each of the coordinate axes, the initial volume will be $dV = dxdydz$.

The final volume in the deformed configuration will have the dimensions dr_1, dr_2, dr_3, and each one will be

$$dr_i = \frac{\partial r_i}{\partial x}dx + \frac{\partial r_i}{\partial y}dy + \frac{\partial r_i}{\partial z}dz \tag{2.2.1.18}$$

If one names $\mathbf{j}_i$ $(i = 1,2,3)$ the vectors that define the volume of the deformed material, their expression in the initial coordinate axes will be

$$\mathbf{j}_1 = \begin{bmatrix} \frac{\partial r_1}{\partial x} & \frac{\partial r_2}{\partial x} & \frac{\partial r_3}{\partial x} \end{bmatrix}^T dx = \mathbf{r}_x dx \tag{2.2.1.19a}$$

$$\mathbf{j}_2 = \begin{bmatrix} \frac{\partial r_1}{\partial y} & \frac{\partial r_2}{\partial y} & \frac{\partial r_3}{\partial y} \end{bmatrix}^T dy = \mathbf{r}_y dy \tag{2.2.1.19b}$$

$$\mathbf{j}_3 = \begin{bmatrix} \frac{\partial r_1}{\partial z} & \frac{\partial r_2}{\partial z} & \frac{\partial r_3}{\partial z} \end{bmatrix}^T dz = \mathbf{r}_z dz \tag{2.2.1.19c}$$

The volume after the body deformation is expressed as

$$dV_r = \mathbf{j}_1 \cdot (\mathbf{j}_2 \times \mathbf{j}_3) = \mathbf{r}_x dx \cdot \left(\mathbf{r}_y dy \times \mathbf{r}_z dz\right) = \left[\mathbf{r}_x \cdot \left(\mathbf{r}_y \times \mathbf{r}_z\right)\right] dxdydz \tag{2.2.1.20}$$

and therefore

$$dV_r = JdV \tag{2.2.1.21}$$

2.2.2 Tensors in small strain theory: Cauchy strain tensor

In the case of small strains, one can disregard the quadratic terms of the strain tensors. Thus, the expressions are simplified and the corresponding tensor is called a *linear strain tensor* or *Cauchy strain tensor* **E**. One can verify that, in this case, disregarding the quadratic terms, one obtains

$$\mathbf{E} = \mathbf{E}_l = \mathbf{E}_e = \frac{1}{2}\left(\mathbf{J}_d + \mathbf{J}_d^T\right) = \frac{1}{2}\left(\mathbf{J} + \mathbf{J}^T - 2\mathbf{I}\right) \quad (2.2.2.1)$$

Recalling the expressions (2.2.1.16) and (2.2.1.17) and eliminating the quadratic terms

$$\mathbf{E} = \begin{bmatrix} \frac{\partial u}{\partial x} & \frac{1}{2}\left(\frac{\partial u}{\partial y} + \frac{\partial v}{\partial x}\right) & \frac{1}{2}\left(\frac{\partial u}{\partial z} + \frac{\partial w}{\partial x}\right) \\ \frac{1}{2}\left(\frac{\partial u}{\partial y} + \frac{\partial v}{\partial x}\right) & \frac{\partial v}{\partial y} & \frac{1}{2}\left(\frac{\partial v}{\partial z} + \frac{\partial w}{\partial y}\right) \\ \frac{1}{2}\left(\frac{\partial u}{\partial z} + \frac{\partial w}{\partial x}\right) & \frac{1}{2}\left(\frac{\partial v}{\partial z} + \frac{\partial w}{\partial y}\right) & \frac{\partial w}{\partial z} \end{bmatrix} = \begin{bmatrix} e_{11} & e_{12} & e_{13} \\ e_{21} & e_{22} & e_{23} \\ e_{31} & e_{32} & e_{33} \end{bmatrix} \quad (2.2.2.2)$$

so that the expressions (2.2.1.5) and (2.2.1.9) are

$$dl_r^2 - dl^2 = 2d\mathbf{x}^T \mathbf{E} d\mathbf{x} \quad (2.2.2.3)$$

In tensor **E**, as in $\mathbf{E}_l$ and $\mathbf{E}_e$, the terms of the main diagonal represent the longitudinal strains in small strains. One can verify this by making the vector $d\mathbf{x}$ coincide with one of the coordinate directions, for example $dl = dx$ and therefore $dy = dz = 0$. Introducing these in (2.2.2.3) gives

$$dl_r^2 - dl^2 = (dl_r - dl)(dl_r + dl) = 2e_{11}dx^2 \quad (2.2.2.4)$$

in small strains

$$dl_r + dl \approx 2dl \quad (2.2.2.5)$$

$$dl_r - dl = e_{11}dx = \frac{\partial u}{\partial x}dx = \varepsilon_x dx \quad (2.2.2.5a)$$

as can be deduced by comparing the previous expression with (2.2.1.17a). The same happens in the others coordinate axes; therefore

$$e_{22} = \varepsilon_y \quad (2.2.2.5b)$$

$$e_{33} = \varepsilon_z \quad (2.2.2.5c)$$

The elements external to the main diagonal are associated with the angular strains. One can verify this by defining an elementary rectangle with sides parallel to the coordinate axes, for example in the directions x, z.

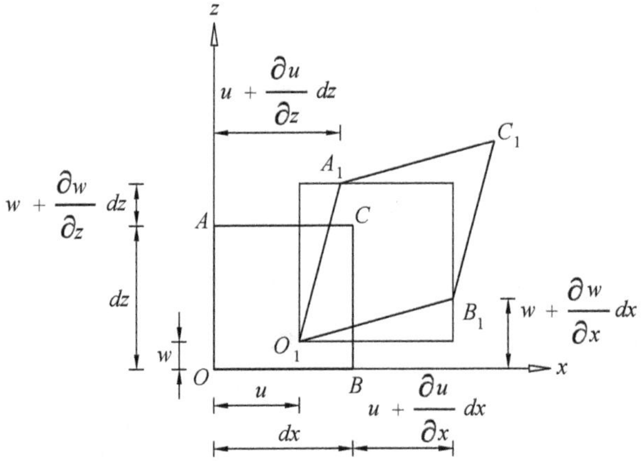

Figure 2.2.2.1: Deformation of an elementary rectangle

The variation of the angle that corresponds to the point O can be expressed only taking into consideration infinitesimals of first order in the following way

$$A\hat{O}B - A_1\hat{O}_1B_1 = \frac{\partial w}{\partial x} + \frac{\partial u}{\partial z} = \gamma_{xz} \tag{2.2.2.6}$$

comparing (2.2.2.6) with the elements of the tensor **E**, one obtains

$$e_{13} = \frac{1}{2} \cdot \left(A\hat{O}B - A_1\hat{O}_1B_1\right) = \frac{1}{2}\gamma_{xz} \tag{2.2.2.7a}$$

The same occurs with the terms

$$e_{12} = \frac{1}{2}\gamma_{xy} \tag{2.2.2.7b}$$

$$e_{23} = \frac{1}{2}\gamma_{yz} \tag{2.2.2.7c}$$

therefore, Cauchy's tensor of small strains can be expressed as

$$\mathbf{E} = \begin{bmatrix} \frac{\partial u}{\partial x} & \frac{1}{2}\left(\frac{\partial u}{\partial y} + \frac{\partial v}{\partial x}\right) & \frac{1}{2}\left(\frac{\partial u}{\partial z} + \frac{\partial w}{\partial x}\right) \\ \frac{1}{2}\left(\frac{\partial u}{\partial y} + \frac{\partial v}{\partial x}\right) & \frac{\partial v}{\partial y} & \frac{1}{2}\left(\frac{\partial v}{\partial z} + \frac{\partial w}{\partial y}\right) \\ \frac{1}{2}\left(\frac{\partial u}{\partial z} + \frac{\partial w}{\partial x}\right) & \frac{1}{2}\left(\frac{\partial v}{\partial z} + \frac{\partial w}{\partial y}\right) & \frac{\partial w}{\partial z} \end{bmatrix} = \begin{bmatrix} \varepsilon_x & \frac{1}{2}\gamma_{xy} & \frac{1}{2}\gamma_{xz} \\ \frac{1}{2}\gamma_{xy} & \varepsilon_y & \frac{1}{2}\gamma_{yz} \\ \frac{1}{2}\gamma_{xz} & \frac{1}{2}\gamma_{yz} & \varepsilon_z \end{bmatrix} \qquad (2.2.2.8)$$

2.2.3 Another definition of strain: Cauchy–Green strain tensor

Previously, the strain has been defined based on the expression $dl_r^2 - dl^2$, according to (2.2.1.4). However, one can define it in other ways, which are the origins of other strain tensors. If, for example, one defines the stretch λ as

$$\lambda = \frac{dl_r}{dl} \qquad (2.2.3.1)$$

one will have

$$\lambda^2 = \frac{dl_r^2}{dl^2} = \frac{\mathbf{dx}^T\mathbf{J}^T\mathbf{J}d\mathbf{x}}{dl^2} = \mathbf{n}^T\mathbf{J}^T\mathbf{J}\mathbf{n} \qquad (2.2.3.2)$$

the product $\mathbf{J}^T\mathbf{J}$ represents a strain tensor named the *right tensor of Cauchy–Green*, $\mathbf{C}_r$

$$\mathbf{C}_r = \mathbf{J}^T\mathbf{J} \qquad (2.2.3.3)$$

This tensor is related to those defined above. Recalling (2.2.1.5)

$$\frac{dl_r^2 - dl^2}{dl^2} = \frac{2\mathbf{dx}^T\mathbf{E}_l\mathbf{dx}}{dl^2} = 2\mathbf{n}^T\mathbf{E}_l\mathbf{n} = \frac{dl_r^2}{dl^2} - 1 = \mathbf{n}^T\mathbf{J}^T\mathbf{J}\mathbf{n} - 1 \qquad (2.2.3.4)$$

comparing with (2.2.3.2), one obtains

$$\mathbf{J}^T\mathbf{J} - \mathbf{I} = \mathbf{C}_r - \mathbf{I} = 2\mathbf{E}_l \qquad (2.2.3.5)$$

then

$$\mathbf{J}^T\mathbf{J} = \mathbf{C}_r = 2\mathbf{E}_l + \mathbf{I} \qquad (2.2.3.6)$$

Furthermore, another strain tensor can be defined similar to this one starting from the Euler–Almansi strain tensor. Recalling (2.2.1.11) and substituting in (2.2.3.6)

$$\mathbf{J}^T\mathbf{J} = 2\mathbf{J}^T\mathbf{E}_e\mathbf{J} + \mathbf{I} \tag{2.2.3.7}$$

multiplying by $\mathbf{J}$ and $\mathbf{J}^T$, one obtains

$$\mathbf{J}\mathbf{J}^T\mathbf{J}\mathbf{J}^T = 2\mathbf{J}\mathbf{J}^T\mathbf{E}_e\mathbf{J}\mathbf{J}^T + \mathbf{J}\mathbf{J}^T \tag{2.2.3.8}$$

The *Cauchy–Green left tensor* is defined as

$$\mathbf{C}_l = \mathbf{J}\mathbf{J}^T \tag{2.2.3.9}$$

substituting in the previous expression

$$\mathbf{C}_l\mathbf{C}_l = 2\mathbf{C}_l\mathbf{E}_e\mathbf{C}_l + \mathbf{C}_l \tag{2.2.3.10}$$

and also

$$\mathbf{C}_l^{-1}\mathbf{C}_l\mathbf{C}_l\mathbf{C}_l^{-1} = 2\mathbf{C}_l^{-1}\mathbf{C}_l\mathbf{E}_e\mathbf{C}_l\mathbf{C}_l^{-1} + \mathbf{C}_l^{-1}\mathbf{C}_l\mathbf{C}_l^{-1} \tag{2.2.3.11}$$

which reduces to

$$\mathbf{I} = 2\mathbf{E}_e + \mathbf{C}_l^{-1} \tag{2.2.3.12}$$

and then

$$\mathbf{E}_e = \frac{1}{2}\left(\mathbf{I} - \mathbf{C}_l^{-1}\right) \tag{2.2.3.13}$$

Both Cauchy–Green tensors are symmetrical, and the following relationship exists between them

$$\mathbf{C}_l = \mathbf{J}^{-1^T}\mathbf{C}_r\mathbf{J}^T \qquad \mathbf{C}_r = \mathbf{J}^{-1}\mathbf{C}_l\mathbf{J} \tag{2.2.3.14}$$

2.3 PRINCIPAL DIRECTIONS OF STRAIN AND MAXIMUM ANGULAR STRAIN

In a generic direction $\mathbf{n}$ of a point of the body, the strain vector $\boldsymbol{\varepsilon}_n$, taking into consideration the Cauchy strain tensor, will be expressed by

$$\boldsymbol{\varepsilon}_n = \begin{bmatrix} \varepsilon_{nx} \\ \varepsilon_{ny} \\ \varepsilon_{nz} \end{bmatrix} = \begin{bmatrix} \varepsilon_x & \frac{1}{2}\gamma_{xy} & \frac{1}{2}\gamma_{xz} \\ \frac{1}{2}\gamma_{xy} & \varepsilon_y & \frac{1}{2}\gamma_{yz} \\ \frac{1}{2}\gamma_{xz} & \frac{1}{2}\gamma_{yz} & \varepsilon_z \end{bmatrix} \begin{bmatrix} l \\ m \\ n \end{bmatrix} = \mathbf{E}\mathbf{n} \tag{2.3.1}$$

l, m, n being the cosine directors of direction $\mathbf{n}$.

The longitudinal component ε of the strain is

$$\varepsilon = \mathbf{n}^T \boldsymbol{\varepsilon}_n = \mathbf{n}^T \mathbf{E}\mathbf{n} \qquad (2.3.2)$$

developing (2.3.2)

$$\varepsilon = \begin{bmatrix} l & m & n \end{bmatrix} \begin{bmatrix} \varepsilon_x & \frac{1}{2}\gamma_{xy} & \frac{1}{2}\gamma_{xz} \\ \frac{1}{2}\gamma_{xy} & \varepsilon_y & \frac{1}{2}\gamma_{yz} \\ \frac{1}{2}\gamma_{xz} & \frac{1}{2}\gamma_{yz} & \varepsilon_z \end{bmatrix} \begin{bmatrix} l \\ m \\ n \end{bmatrix} \qquad (2.3.3)$$

which leads to

$$\varepsilon = \varepsilon_x l^2 + \varepsilon_y m^2 + \varepsilon_z n^2 + \gamma_{xy} lm + \gamma_{xz} ln + \gamma_{yz} mn \qquad (2.3.4)$$

The angular strain is

$$\frac{1}{2}\gamma = \sqrt{|\boldsymbol{\varepsilon}_n|^2 - \varepsilon^2} \qquad (2.3.5)$$

The *principal directions of strain* are those in which only the longitudinal strain exists because the angular strain is cancelled out. Therefore, in those directions the strain vector $\boldsymbol{\varepsilon}_n$ is parallel to vector $\mathbf{n}$

$$\boldsymbol{\varepsilon}_n = \mathbf{E}\mathbf{n} = \varepsilon \mathbf{n} \qquad (2.3.6)$$

which is equivalent to the expression

$$(\mathbf{E} - \varepsilon \mathbf{I})\,\mathbf{n} = \mathbf{0} \qquad (2.3.7)$$

as a result, the problem consists in obtaining the eigenvalues of (2.3.7). That leads to a third degree equation

$$\varepsilon^3 - I_1 \varepsilon^2 + I_2 \varepsilon - I_3 = 0 \qquad (2.3.8)$$

where I_1, I_2, I_3 have the same value in any system of Cartesian coordinates. One refers to them as *invariants of the strain tensor*, and they are

linear invariant $\quad I_1 = \varepsilon_x + \varepsilon_y + \varepsilon_z \qquad$ (2.3.9a)

quadratic invariant $\quad I_2 = \varepsilon_x \varepsilon_y + \varepsilon_x \varepsilon_z + \varepsilon_y \varepsilon_z - \frac{1}{4}\gamma_{xy}^2 - \frac{1}{4}\gamma_{xz}^2 - \frac{1}{4}\gamma_{yz}^2 \qquad$ (2.3.9b)

cubic invariant $\quad I_3 = |\mathbf{E}| \qquad$ (2.3.9c)

The roots of (2.3.8) are always real, are known as *principal strains* and are

$$\varepsilon_1 \geq \varepsilon_2 \geq \varepsilon_3 \tag{2.3.10}$$

The principal direction associated with each one is obtained by substituting the value ε_i $(i = 1, 2, 3)$ of each principal strain in the following system of equations

$$\varepsilon_x l + \frac{1}{2}\gamma_{xy} m + \frac{1}{2}\gamma_{xz} n - \varepsilon_i l = 0 \tag{2.3.11a}$$

$$\frac{1}{2}\gamma_{xy} l + \varepsilon_y m + \frac{1}{2}\gamma_{yz} n - \varepsilon_i m = 0 \tag{2.3.11b}$$

$$\frac{1}{2}\gamma_{xz} l + \frac{1}{2}\gamma_{yz} m + \varepsilon_z n - \varepsilon_i n = 0 \tag{2.3.11c}$$

$$l^2 + m^2 + n^2 = 1 \tag{2.3.11d}$$

As there is no angular strain in the principal directions, the strain tensor **E** has the expression

$$\mathbf{E} = \begin{bmatrix} \varepsilon_1 & 0 & 0 \\ 0 & \varepsilon_2 & 0 \\ 0 & 0 & \varepsilon_3 \end{bmatrix} \tag{2.3.12}$$

From this, recalling equation (2.3.5) and using the principal directions of strain as coordinate axes, the angular strain in any direction can be written as:

$$\left(\frac{1}{2}\gamma\right)^2 = \varepsilon_1^2 l^2 + \varepsilon_2^2 m^2 + \varepsilon_3^2 n^2 - \left(\varepsilon_1 l^2 + \varepsilon_2 m^2 + \varepsilon_3 n^2\right)^2 \tag{2.3.13}$$

As the components of **n** must meet condition (2.3.11d), the maximum value of γ is obtained by solving a constrained maximization problem that can be carried out using the method of Lagrange multipliers, creating a Lagrangian formulation with equation (2.3.13) of the strain that one wants to maximize and condition (2.3.11d)

$$L = \varepsilon_1^2 l^2 + \varepsilon_2^2 m^2 + \varepsilon_3^2 n^2 - \left(\varepsilon_1 l^2 + \varepsilon_2 m^2 + \varepsilon_3 n^2\right)^2 + \lambda\left(l^2 + m^2 + n^2 - 1\right) \tag{2.3.14}$$

The extreme values of (2.3.14) shall comply with the conditions

$$\frac{\partial L}{\partial l} = \varepsilon_1^2 l - 2\varepsilon_1\left(\varepsilon_1 l^2 + \varepsilon_2 m^2 + \varepsilon_3 n^2\right) l + \lambda l = 0 \tag{2.3.15a}$$

$$\frac{\partial L}{\partial m} = \varepsilon_2^2 m - 2\varepsilon_2\left(\varepsilon_1 l^2 + \varepsilon_2 m^2 + \varepsilon_3 n^2\right) m + \lambda m = 0 \tag{2.3.15b}$$

$$\frac{\partial L}{\partial n} = \varepsilon_3^2 n - 2\varepsilon_3\left(\varepsilon_1 l^2 + \varepsilon_2 m^2 + \varepsilon_3 n^2\right) n + \lambda n = 0 \tag{2.3.15c}$$

$$\frac{\partial L}{\partial \lambda} = l^2 + m^2 + n^2 - 1 = 0 \tag{2.3.15d}$$

Equations (2.3.15) constitute a non-linear system. The maximum value of the Lagrangian, which is also the maximum angular strain, is the one in which $m = 0$. If this value is substituted in the system, equations (2.3.15a) and (2.3.15c) are simplified, and expressing the value of λ in each of them gives

$$\varepsilon_1\left[\varepsilon_1 - 2\left(\varepsilon_1 l^2 + \varepsilon_3 n^2\right)\right] = \varepsilon_3\left[\varepsilon_3 - 2\left(\varepsilon_1 l^2 + \varepsilon_3 n^2\right)\right] \tag{2.3.16}$$

grouping terms and dividing by $(\varepsilon_1 - \varepsilon_3)$ results in

$$\varepsilon_1 + \varepsilon_3 - 2\left(\varepsilon_1 l^2 + \varepsilon_3 n^2\right) = 0 \tag{2.3.17}$$

this equation, together with (2.3.15d), provides the solution

$$m = 0 \tag{2.3.18a}$$

$$l^2 = n^2 = 0.5 \tag{2.3.18b}$$

One can verify that the planes associated with the directions of (2.3.18) form angles of 45° with the principal directions ε_1, ε_3 and contain the one with the direction ε_2.

Substituting into equation (2.3.13), one obtains the maximum value of the angular strain

$$\frac{1}{2}\gamma_{max} = \frac{\varepsilon_1 - \varepsilon_3}{2} \tag{2.3.19}$$

If one considers an elementary prism of dimensions *dx*, *dy*, *dz*, the increase in volume experienced as a result of the strain will be

$$\frac{dV}{V} = \frac{dx\left(1 + \varepsilon_x\right) dy\left(1 + \varepsilon_y\right) dz\left(1 + \varepsilon_z\right) - dxdydz}{dxdydz} \approx \varepsilon_x + \varepsilon_y + \varepsilon_z = e \tag{2.3.20}$$

parameter e is known as the *volumetric strain*, and it is evident that it coincides with the invariant I_1 according to equation (2.3.9a); thus, its value is constant and independent of the coordinate system chosen.

If the strain tensor is the Green–Lagrange, then the expression to obtain the principal strains is

$$|\mathbf{E}_l - \varepsilon \mathbf{I}| = 0 \qquad (2.3.21)$$

which provides three roots $\varepsilon_i \ (i = 1,2,3)$. The principal directions are obtained from the following equation systems

$$(\mathbf{E}_l - \varepsilon_i \mathbf{I})\,\mathbf{n}_i = \mathbf{0} \quad i = 1,2,3 \qquad (2.3.22)$$

if in equation (2.3.21) one introduces (2.2.3.6), which relates the Green–Lagrange tensor and the Cauchy–Green right tensor, this results in

$$|\mathbf{E}_l - \varepsilon \mathbf{I}| = \left|\left(\frac{\mathbf{C}_r}{2} - \frac{\mathbf{I}}{2}\right) - \varepsilon \mathbf{I}\right| = 0 \qquad (2.3.23)$$

or

$$|\mathbf{C}_r - (2\varepsilon + 1)\,\mathbf{I}| = 0 \qquad (2.3.24)$$

which means that the eigenvalues of the Green–Lagrange tensor can be obtained from equation (2.3.21) or (2.3.24). Therefore, the directions of the principal strain for the Green–Lagrange tensor and the Cauchy–Green right tensor are the same.

This last expression can be written as

$$\left|\mathbf{C}_r - \lambda^2 \mathbf{I}\right| = 0 \qquad (2.3.25)$$

and it represents the way of obtaining the eigenvalues of the Cauchy–Green right tensor. Comparing (2.3.24) and (2.3.25), it is clear that the eigenvalues of $\mathbf{C}_r$ and $\mathbf{E}_l$ are related in the form

$$\lambda_i = \sqrt{1 + 2\varepsilon_i} \quad i = 1,2,3 \qquad (2.3.26)$$

a linear approximation is used occasionally for this expression

$$\lambda_i \approx 1 + \varepsilon_i \quad i = 1,2,3 \qquad (2.3.27)$$

The reason that the eigenvalues of the Cauchy–Green right tensor are written as $\lambda_i^2 \ (i = 1,2,3)$ is because they are the values of the principal strains of this tensor. One may recall that they are defined by the expressions $\mathbf{n}_i^T \mathbf{C}_r \mathbf{n}_i \ (i = 1,2,3)$. If one substitutes $\mathbf{C}_r$ for its expression in (2.2.3.3), the result is

$$\mathbf{n}_i^T \mathbf{C}_r \mathbf{n}_i = \mathbf{n}_i^T \mathbf{J}^T \mathbf{J} \mathbf{n}_i \quad i = 1,2,3 \qquad (2.3.28)$$

and, comparing this with (2.2.3.2), it finally becomes

$$\mathbf{n}_i^T \mathbf{C}_r \mathbf{n}_i = \mathbf{n}_i^T \mathbf{J}^T \mathbf{J} \mathbf{n}_i = \lambda_i^2 \quad i = 1,2,3 \qquad (2.3.29)$$

If the coordinate axes are the principal directions of strain, the expressions of the invariants of the tensor $\mathbf{C}_r$ are

$$I_{C1} = \lambda_1^2 + \lambda_2^2 + \lambda_3^2 \tag{2.3.30a}$$

$$I_{C2} = \lambda_1^2\lambda_2^2 + \lambda_1^2\lambda_3^2 + \lambda_2^2\lambda_3^2 \tag{2.3.30b}$$

$$I_{C3} = \lambda_1^2\lambda_2^2\lambda_3^2 \tag{2.3.30c}$$

and the Cauchy–Green right tensor in the axes corresponding to the principal directions will be

$$\mathbf{C}_r = \begin{bmatrix} \lambda_1^2 & 0 & 0 \\ 0 & \lambda_2^2 & 0 \\ 0 & 0 & \lambda_3^2 \end{bmatrix} \tag{2.3.31}$$

Also, in these axes, the volume variation of the deformable body with respect to the initial one is

$$\frac{V_r}{V} \approx \frac{dxdydz\,(1+\varepsilon_1)\,(1+\varepsilon_2)\,(1+\varepsilon_3)}{dxdydz} = \lambda_1\lambda_2\lambda_3 = \sqrt{I_{C3}} \tag{2.3.32}$$

The principal strains of the Cauchy–Green left tensor $\mathbf{C}_l$ are obtained from the equation

$$\left|\mathbf{C}_l - \mu^2\mathbf{I}\right| = 0 \tag{2.3.33}$$

One can demonstrate that the eigenvalues and, consequently, the invariants of the Cauchy–Green left tensor $\mathbf{C}_l$ are the same as the right tensor, and therefore are also called I_{Ci} $(i = 1,2,3)$. The principal directions of $\mathbf{C}_l$ are obtained by rotating the coordinates system formed by the principal directions of $\mathbf{C}_r$.

On occasions, it may be convenient to change the coordinate axes to express the strains of the points of a body. If the current system is replaced by another whose transformation matrix $\mathbf{T}$ is

$$\mathbf{T} = \begin{bmatrix} \cos\alpha_{Xx} & \cos\beta_{Xy} & \cos\gamma_{Xz} \\ \cos\alpha_{Yx} & \cos\beta_{Yy} & \cos\gamma_{Yz} \\ \cos\alpha_{Zx} & \cos\beta_{Zy} & \cos\gamma_{Zz} \end{bmatrix} = \begin{bmatrix} l_X & m_X & n_X \\ l_Y & m_Y & n_Y \\ l_Z & m_Z & n_Z \end{bmatrix} \tag{2.3.34}$$

where

$\alpha_{Xx}, \alpha_{Yx}, \alpha_{Zx}$: angles of the new coordinate system with the x axis
$\beta_{Xy}, \beta_{Yy}, \beta_{Zy}$: angles of the new coordinate system with the y axis
$\gamma_{Xz}, \gamma_{Yz}, \gamma_{Zz}$: angles of the new coordinate system with the z axis

The direction vector $\mathbf{n}$ and the strain vector $\boldsymbol{\varepsilon}_n$ in the new axes will be

$$\bar{\mathbf{n}} = \mathbf{T}\mathbf{n} \quad \bar{\boldsymbol{\varepsilon}}_n = \mathbf{T}\boldsymbol{\varepsilon}_n \tag{2.3.35}$$

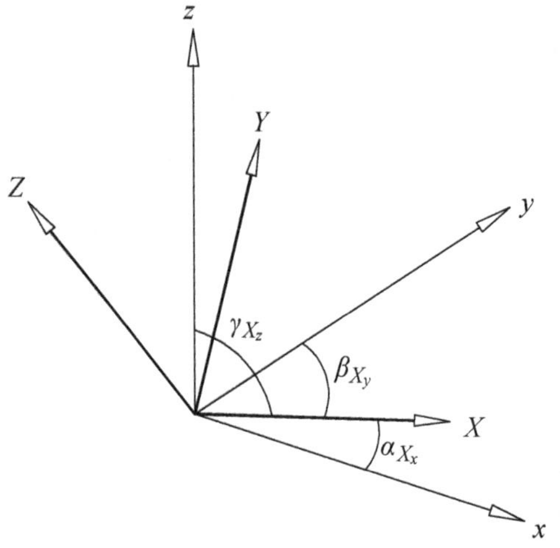

Figure 2.3.1: Change of coordinate system

Assuming that $\boldsymbol{\varepsilon}$ represents any of the strain tensors that have been defined so far, one obtains

$$\overline{\boldsymbol{\varepsilon}}_n = \mathbf{T}\boldsymbol{\varepsilon}\mathbf{n} = \mathbf{T}\boldsymbol{\varepsilon}\mathbf{T}^{-1}\overline{\mathbf{n}} \tag{2.3.36}$$

in the matrix $\mathbf{T}$, which is orthogonal, one has

$$\mathbf{T}^{-1} = \mathbf{T}^T \tag{2.3.37}$$

then

$$\overline{\boldsymbol{\varepsilon}}_n = \mathbf{T}\boldsymbol{\varepsilon}\mathbf{T}^T\overline{\mathbf{n}} = \overline{\boldsymbol{\varepsilon}}\overline{\mathbf{n}} \tag{2.3.38}$$

therefore, the strain tensor in the new coordinates system will be

$$\overline{\boldsymbol{\varepsilon}} = \mathbf{T}\boldsymbol{\varepsilon}\mathbf{T}^T \tag{2.3.39}$$

2.4 ELLIPSOID OF STRAINS

This is the locus of the ends of the strain vector $\boldsymbol{\varepsilon}_n$ corresponding to all the directions defined from a point. If one takes the principal directions as coordinate axes, the expression for the strain at any point is

$$\boldsymbol{\varepsilon}_n = \begin{bmatrix} \varepsilon_{nx} \\ \varepsilon_{ny} \\ \varepsilon_{nz} \end{bmatrix} = \begin{bmatrix} \varepsilon_1 & 0 & 0 \\ 0 & \varepsilon_2 & 0 \\ 0 & 0 & \varepsilon_3 \end{bmatrix} \begin{bmatrix} l \\ m \\ n \end{bmatrix} \tag{2.4.1}$$

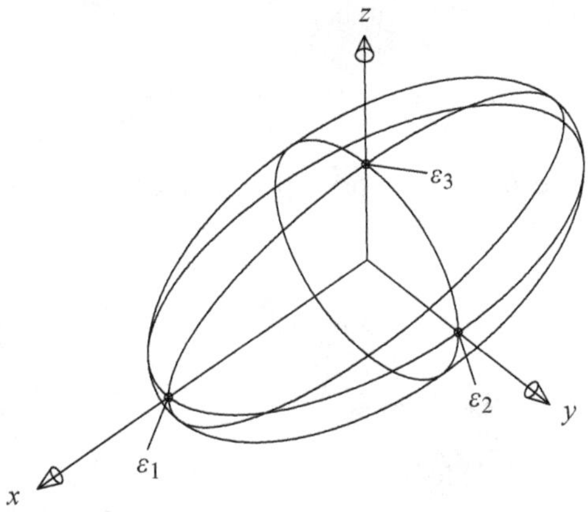

Figure 2.4.1: Ellipsoid of strains

and also

$$l^2 + m^2 + n^2 = 1 \tag{2.4.2}$$

eliminating *l*, *m*, *n* in the previous equations gives

$$\frac{\varepsilon_{nx}^2}{\varepsilon_1^2} + \frac{\varepsilon_{ny}^2}{\varepsilon_2^2} + \frac{\varepsilon_{nz}^2}{\varepsilon_3^2} = 1 \tag{2.4.3}$$

which is the equation of an ellipsoid and is usually written as

$$\frac{x^2}{\varepsilon_1^2} + \frac{y^2}{\varepsilon_2^2} + \frac{z^2}{\varepsilon_3^2} = 1 \tag{2.4.4}$$

This surface only gives an idea of the spatial distribution of the strain vector $\boldsymbol{\varepsilon}_n$, but does not indicate anything with regard to the sign of the vector or its direction. To obtain this information, one should use the equations of other surfaces, called indicatrix quadric and directives quadric.

2.5 INDICATRIX QUADRIC OF STRAINS

This is the locus of the ends of the segments that are defined when taking a length *ON* along each direction **n** from the point *O*, the origin of coordinates.

$$ON = \frac{1}{\sqrt{|\varepsilon|}} \tag{2.5.1}$$

the coordinates of the point N will be

$$x = \frac{l}{\sqrt{|\varepsilon|}} \quad y = \frac{m}{\sqrt{|\varepsilon|}} \quad z = \frac{n}{\sqrt{|\varepsilon|}} \tag{2.5.2}$$

and if the coordinate axes are the principal directions, one obtains

$$\varepsilon = \varepsilon_1 l^2 + \varepsilon_2 m^2 + \varepsilon_3 n^2 \tag{2.5.3}$$

eliminating l, m, n in (2.5.2) and (2.5.3) gives

$$\varepsilon_1 x^2 + \varepsilon_2 y^2 + \varepsilon_3 z^2 = \frac{\varepsilon}{|\varepsilon|} = c = \pm 1 \tag{2.5.4}$$

a) If $\varepsilon_i > 0\ (i = 1, 2, 3)$ for $c = -1$, equation (2.5.4) is an imaginary ellipsoid that has no physical meaning. For $c = 1$, a real ellipsoid is obtained that is the locus of the points ON (figure 2.5.1). All the normal strains are positive and the strain vector $\boldsymbol{\varepsilon}_n$ forms an acute angle with the direction **n**.
b) If $\varepsilon_i < 0\ (i = 1, 2, 3)$, the real ellipsoid appears for $c = -1$. All the normal strains are negative and the total strain $\boldsymbol{\varepsilon}_n$ forms an obtuse angle with the direction **n**.
c) If the two principal strains are positive, $\varepsilon_1 \geq \varepsilon_2 > 0$ and $\varepsilon_3 < 0$, for $c = 1$, a hyperboloid of one sheet is the result, and for c = -1, one obtains a hyperboloid of two sheets, as shown in figure 2.5.2.

   If direction **n** cuts the hyperboloid of one sheet, the normal strain is positive, and when it cuts the hyperboloid of two sheets, it is negative. Both hyperboloids are separated by the asymptote cone that has the equation

$$\varepsilon_1 x^2 + \varepsilon_2 y^2 + \varepsilon_3 z^2 = 0 \tag{2.5.5}$$

   If direction **n** coincides with one of the generatrices of the cone, the normal strain is equal to zero.
d) If a principal strain is positive, $\varepsilon_1 > 0$, and the other two negative, $\varepsilon_3 \leq \varepsilon_2 < 0$, for $c = -1$, a hyperboloid of one sheet is obtained; for $c = 1$, a hyperboloid of two sheets appears and the conclusions are similar to those of the previous case.

The indicatrix quadric enables one to determine the total strain corresponding to a direction **n**. Figures 2.5.3 and 2.5.4 indicate how to obtain $\boldsymbol{\varepsilon}_n$ from **n** in the ellipsoid and the hyperboloid of strain.

The graphic explanation of figures 2.5.3 and 2.5.4 can be obtained analytically. If the equation of the indicatrix quadric is

$$f(x, y, z) = \varepsilon_1 x^2 + \varepsilon_2 y^2 + \varepsilon_3 z^2 - c \tag{2.5.6}$$

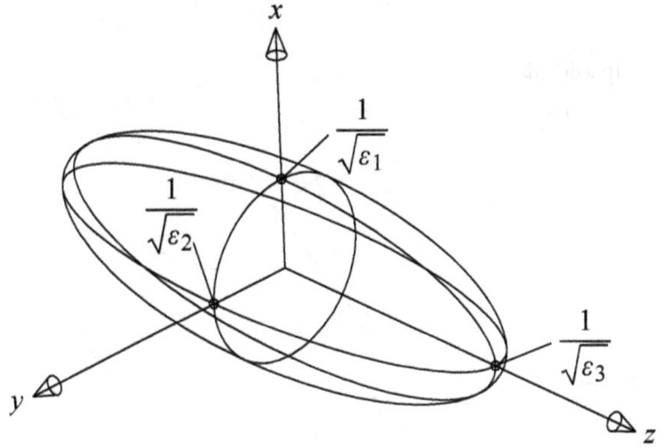

Figure 2.5.1: Indicatrix ellipsoid of strain

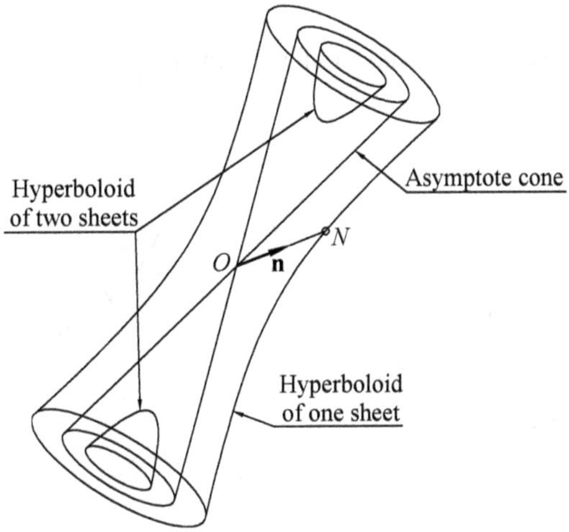

Figure 2.5.2: Indicatrix hyperboloid of strain

the conjugate plane of a direction **n** is

$$x\frac{\partial f}{\partial x} + y\frac{\partial f}{\partial y} + z\frac{\partial f}{\partial z} = 0 \tag{2.5.7}$$

where

$$\frac{\partial f}{\partial x} = 2\varepsilon_1 x \quad \frac{\partial f}{\partial y} = 2\varepsilon_2 y \quad \frac{\partial f}{\partial z} = 2\varepsilon_3 z \tag{2.5.8}$$

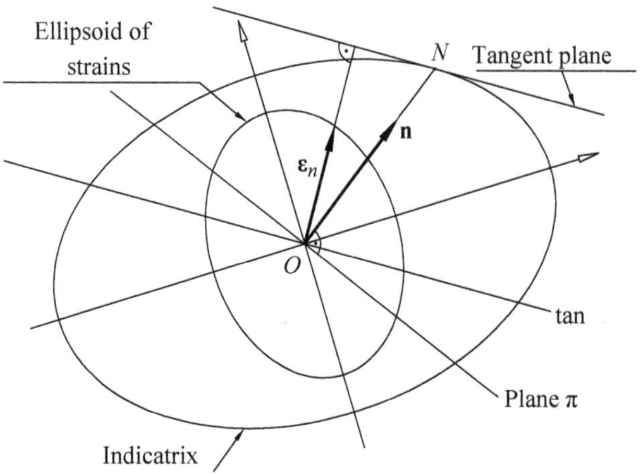

Figure 2.5.3: Direction **n** and strain $\boldsymbol{\varepsilon}_n$ in the indicatrix ellipsoid of strain

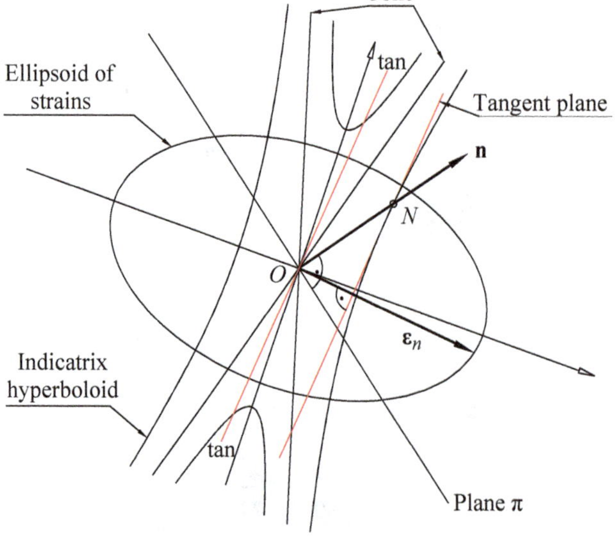

Figure 2.5.4: Direction **n** and strain $\boldsymbol{\varepsilon}_n$ in the indicatrix hyperboloid of strain

substituting x, y, z in (2.5.8) with l, m, n, equation (2.5.7), which defines the conjugate plane, becomes

$$l\varepsilon_1 x + m\varepsilon_2 y + n\varepsilon_3 z = 0 \tag{2.5.9}$$

One observes that the components $(l\varepsilon_1, m\varepsilon_2, n\varepsilon_3)$ of the vector **n** that defines the plane are also the components of the total strain $\boldsymbol{\varepsilon}_n$, which correspond to the direction **n** when the coordinate axes are the principal directions, as defined in the equation (2.4.1). Therefore, the perpendicular vector to that plane, which goes from the origin of coordinates to the ellipsoid of strains, is $\boldsymbol{\varepsilon}_n$.

2.6 DIRECTRIX QUADRIC OF STRAINS

This is the locus defined by the points N resulting from changing the strain vector $\boldsymbol{\varepsilon}_n$ in the following way

$$\frac{\boldsymbol{\varepsilon}_n}{\sqrt{|\varepsilon|}} \tag{2.6.1}$$

If the coordinate axes are the principal directions, the strain components that correspond to a direction **n** are

$$\varepsilon_{nx} = \varepsilon_1 l \quad \varepsilon_{ny} = \varepsilon_2 m \quad \varepsilon_{nz} = \varepsilon_3 n \tag{2.6.2}$$

the coordinates of the point N are

$$x = \frac{\varepsilon_1 l}{\sqrt{|\varepsilon|}} \quad y = \frac{\varepsilon_2 m}{\sqrt{|\varepsilon|}} \quad z = \frac{\varepsilon_3 n}{\sqrt{|\varepsilon|}} \tag{2.6.3}$$

and, taking into consideration equation (2.5.3), gives

$$\frac{x^2}{\varepsilon_1} + \frac{y^2}{\varepsilon_2} + \frac{z^2}{\varepsilon_3} = \pm 1 \tag{2.6.4}$$

One can discuss the equation of this quadric in the same way as the indicatrix quadric.

This quadric has the attribute that, by knowing the direction of the strain vector $\boldsymbol{\varepsilon}_n$, which provides point N in figure 2.6.2 in the directrix quadric, the direction **n** to which this strain corresponds is the normal direction to the tangent plane in the intersection of $\boldsymbol{\varepsilon}_n$ with the directrix quadric, as shown in this figure.

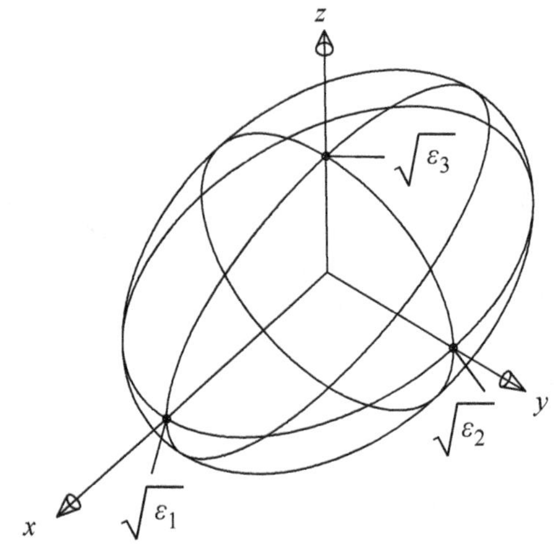

Figure 2.6.1: Directrix ellipsoid of strains

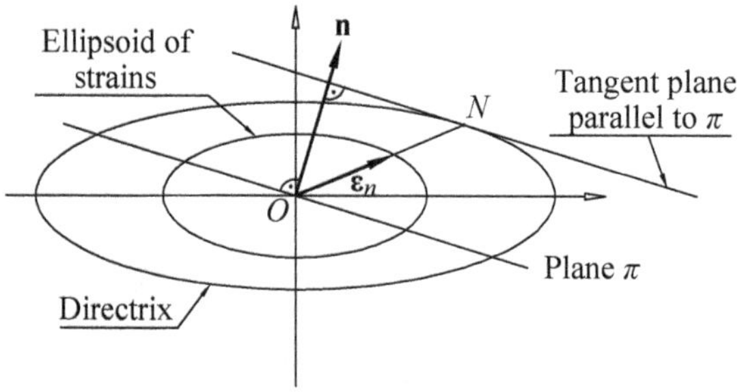

Figure 2.6.2: Plane section of the ellipsoid quadrix of strains

2.7 MOHR'S CIRCLES OF STRAIN

This is an approach that studies the strain tensor of any direction **n**. Recalling equation (2.4.1), one obtains

$$\boldsymbol{\varepsilon}_n^T \boldsymbol{\varepsilon}_n = |\boldsymbol{\varepsilon}_n|^2 = (\varepsilon_1 l)^2 + (\varepsilon_2 m)^2 + (\varepsilon_3 n)^2 = \varepsilon^2 + \left(\frac{1}{2}\gamma\right)^2 \tag{2.7.1}$$

also

$$\varepsilon = \mathbf{n}^T \boldsymbol{\varepsilon}_n = \varepsilon_1 l^2 + \varepsilon_2 m^2 + \varepsilon_3 n^2 \tag{2.7.2}$$

and

$$l^2 + m^2 + n^2 = 1 \tag{2.7.3}$$

eliminating m and n in the three equations results in

$$\begin{vmatrix} \varepsilon^2 + \frac{1}{4}\gamma^2 - \varepsilon_1^2 l^2 & \varepsilon_2^2 & \varepsilon_3^2 \\ \varepsilon - \varepsilon_1 \cdot l^2 & \varepsilon_2 & \varepsilon_3 \\ 1 - l^2 & 1 & 1 \end{vmatrix} = 0 \quad (2.7.4)$$

developing and dividing by $(\varepsilon_2 - \varepsilon_3)$ and recalling that $\varepsilon_1 \geq \varepsilon_2 \geq \varepsilon_3$, one obtains the following expression

$$\frac{1}{4}\gamma^2 + \left(\varepsilon - \frac{\varepsilon_2 + \varepsilon_3}{2}\right)^2 - \left(\frac{\varepsilon_2 - \varepsilon_3}{2}\right)^2 = l^2 (\varepsilon_1 - \varepsilon_2)(\varepsilon_1 - \varepsilon_3) \geq 0 \quad (2.7.5a)$$

Similarly, eliminating l and n gives

$$\frac{1}{4}\gamma^2 + \left(\varepsilon - \frac{\varepsilon_1 + \varepsilon_3}{2}\right)^2 - \left(\frac{\varepsilon_1 - \varepsilon_3}{2}\right)^2 = -m^2 (\varepsilon_2 - \varepsilon_3)(\varepsilon_1 - \varepsilon_2) \leq 0 \quad (2.7.5b)$$

and eliminating l and m gives

$$\frac{1}{4}\gamma^2 + \left(\varepsilon - \frac{\varepsilon_1 + \varepsilon_2}{2}\right)^2 - \left(\frac{\varepsilon_1 - \varepsilon_2}{2}\right)^2 = n^2 (\varepsilon_3 - \varepsilon_1)(\varepsilon_3 - \varepsilon_2) \geq 0 \quad (2.7.5c)$$

The locus defined by equations (2.7.5) is the set of points located within the circle defined by (2.7.5b) and outside the circles defined by (2.7.5a) and (2.7.5c). This is shown in figure 2.7.1, and these points represent the set of strain vectors of all **n** directions. If the strain of a point is defined by ε and $\gamma/2$, the point $(\varepsilon, \gamma/2)$ represents the end of the strain vector $\boldsymbol{\varepsilon}_n$.

Mohr's circles can identify the point that defines the strain vector for any direction **n** defined with regard to the principal directions of strain as $\mathbf{n} = l\mathbf{i} + m\mathbf{j} + n\mathbf{k}$. One remembers that the outer circle was obtained by eliminating l and n in equations (2.7.1), (2.7.2) and (2.7.3). If one defines

$$l = \cos\alpha \qquad n = \cos\theta \quad (2.7.6)$$

and the angles α and θ are defined from the ε axis, one obtains the points P and Q, as shown in figure 2.7.2. If A and B are the centres of the inner circles, drawing from the ε axis, the circles of radius AP and BQ, respectively, one obtains the point M, which represents the strain for that direction.

This graphical construction can also verify that the maximum angular strains correspond to points M_1 and M_2 and that they are obtained with $\alpha = \theta = 45°$, as noted in figure 2.7.3. This means that

$$l = \cos\alpha = \pm\frac{\sqrt{2}}{2} \qquad n = \cos\theta = \pm\frac{\sqrt{2}}{2} \quad (2.7.7)$$

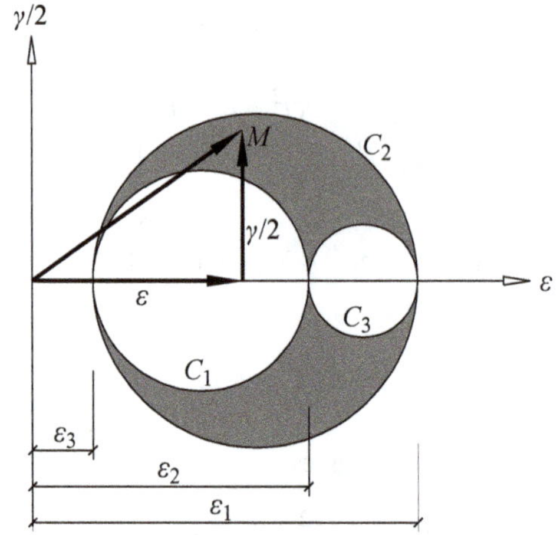

Figure 2.7.1: Mohr's circles of strain

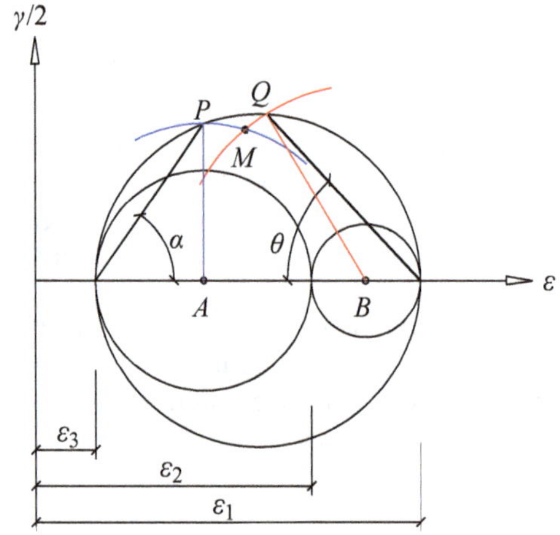

Figure 2.7.2: Graphical calculation of the strain vector

and therefore $m = 0$, which is what was analytically obtained in section 2.3. Also, one can verify that the maximum value obtained there was

$$\frac{1}{2}\gamma_{max} = \frac{\varepsilon_1 - \varepsilon_3}{2} \tag{2.7.8}$$

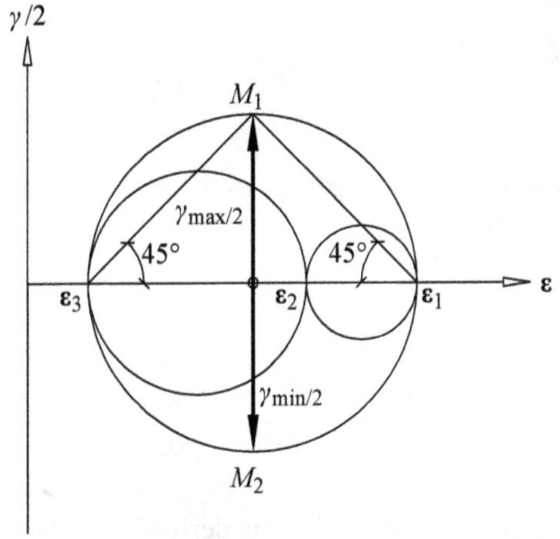

Figure 2.7.3: Directions of maximum angular strain

2.8 COMPATIBILITY EQUATIONS IN THEORY OF SMALL STRAINS

Given a field of displacements defined by the values u, v, w, one can always obtain the field of strains. However, the reverse situation is not always possible because the components of the strain tensor are related to each other as shown below.

From the system of differential equations generated by the strain tensor in linear theory, it is possible to obtain the expressions for the displacements u, v, w of any point on the body.

$$\varepsilon_x = \frac{\partial u}{\partial x} \quad \varepsilon_y = \frac{\partial v}{\partial y} \quad \varepsilon_z = \frac{\partial w}{\partial z} \tag{2.8.1a}$$

$$\gamma_{xy} = \frac{\partial u}{\partial y} + \frac{\partial v}{\partial x} \quad \gamma_{xz} = \frac{\partial u}{\partial z} + \frac{\partial w}{\partial x} \quad \gamma_{yz} = \frac{\partial v}{\partial z} + \frac{\partial w}{\partial y} \tag{2.8.1b}$$

This system of six equations contains only three unknowns; thus, it can only be solved when certain relationships exist between them. That is to say, the components of the strain tensor must meet certain conditions. This is due to the physical requirement to avoid discontinuities or superposition in the material points of the body after the deformation, in other words, the need for the elementary particles to remain united.

If in the components ε_x, ε_y, γ_{xy} the following derivatives are defined

$$\frac{\partial^2 \varepsilon_x}{\partial y^2} = \frac{\partial^3 u}{\partial x \partial y^2} \quad \frac{\partial^2 \varepsilon_y}{\partial x^2} = \frac{\partial^3 v}{\partial x^2 \partial y} \quad \frac{\partial^2 \gamma_{xy}}{\partial x \partial y} = \frac{\partial^3 u}{\partial x \partial y^2} + \frac{\partial^3 v}{\partial x^2 \partial y} \tag{2.8.2}$$

one obtains

$$\frac{\partial^2 \varepsilon_x}{\partial y^2} + \frac{\partial^2 \varepsilon_y}{\partial x^2} = \frac{\partial^2 \gamma_{xy}}{\partial x \partial y} \tag{2.8.3}$$

and, carrying out similar operations with ε_x, ε_z, γ_{xz} and ε_y, ε_z, γ_{yz}, one obtains the following expressions

$$\frac{\partial^2 \varepsilon_x}{\partial z^2} + \frac{\partial^2 \varepsilon_z}{\partial x^2} = \frac{\partial^2 \gamma_{xz}}{\partial x \partial z} \tag{2.8.4a}$$

$$\frac{\partial^2 \varepsilon_y}{\partial z^2} + \frac{\partial^2 \varepsilon_z}{\partial y^2} = \frac{\partial^2 \gamma_{yz}}{\partial y \partial z} \tag{2.8.4b}$$

Also, if in ε_x, γ_{xy}, γ_{xz}, γ_{yz} the following derivatives are carried out

$$\frac{\partial^2 \varepsilon_x}{\partial y \partial z} = \frac{\partial^3 u}{\partial x \partial y \partial z} \tag{2.8.5a}$$

$$\frac{\partial^2 \gamma_{xy}}{\partial x \partial z} = \frac{\partial^3 u}{\partial x \partial y \partial z} + \frac{\partial^3 v}{\partial x^2 \partial z} \quad \frac{\partial^2 \gamma_{xz}}{\partial x \partial y} = \frac{\partial^3 u}{\partial x \partial y \partial z} + \frac{\partial^3 w}{\partial x^2 \partial y}$$

$$\frac{\partial^2 \gamma_{yz}}{\partial x^2} = \frac{\partial^3 v}{\partial x^2 \partial z} + \frac{\partial^3 w}{\partial y \partial x^2} \tag{2.8.5b}$$

this results in

$$2\frac{\partial^2 \varepsilon_x}{\partial y \partial z} = \frac{\partial}{\partial x}\left(\frac{\partial \gamma_{xy}}{\partial z} + \frac{\partial \gamma_{xz}}{\partial y} - \frac{\partial \gamma_{yz}}{\partial x}\right) \tag{2.8.6}$$

Proceeding similarly with ε_y, γ_{xy}, γ_{xz}, γ_{yz}, one obtains

$$2\frac{\partial^2 \varepsilon_y}{\partial y \partial z} = \frac{\partial}{\partial y}\left(\frac{\partial \gamma_{xy}}{\partial z} - \frac{\partial \gamma_{xz}}{\partial y} + \frac{\partial \gamma_{yz}}{\partial x}\right) \tag{2.8.7}$$

and doing the same with ε_z, γ_{xy}, γ_{xz}, γ_{yz} gives

$$2\frac{\partial^2 \varepsilon_z}{\partial x \partial y} = \frac{\partial}{\partial z}\left(-\frac{\partial \gamma_{xy}}{\partial z} + \frac{\partial \gamma_{xz}}{\partial y} + \frac{\partial \gamma_{yz}}{\partial x}\right) \tag{2.8.8}$$

The set of expressions obtained are

$$\frac{\partial^2 \varepsilon_x}{\partial y^2} + \frac{\partial^2 \varepsilon_y}{\partial x^2} = \frac{\partial^2 \gamma_{xy}}{\partial x \partial y} \qquad 2\frac{\partial^2 \varepsilon_x}{\partial y \partial z} = \frac{\partial}{\partial x}\left(\frac{\partial \gamma_{xy}}{\partial z} + \frac{\partial \gamma_{xz}}{\partial y} - \frac{\partial \gamma_{yz}}{\partial x}\right) \tag{2.8.9a}$$

$$\frac{\partial^2 \varepsilon_x}{\partial z^2} + \frac{\partial^2 \varepsilon_z}{\partial x^2} = \frac{\partial^2 \gamma_{xz}}{\partial x \partial z} \qquad 2\frac{\partial^2 \varepsilon_y}{\partial x \partial z} = \frac{\partial}{\partial y}\left(\frac{\partial \gamma_{xy}}{\partial z} - \frac{\partial \gamma_{xz}}{\partial y} + \frac{\partial \gamma_{yz}}{\partial x}\right) \tag{2.8.9b}$$

$$\frac{\partial^2 \varepsilon_y}{\partial z^2} + \frac{\partial^2 \varepsilon_z}{\partial y^2} = \frac{\partial^2 \gamma_{yz}}{\partial y \partial z} \qquad 2\frac{\partial^2 \varepsilon_z}{\partial x \partial y} = \frac{\partial}{\partial z}\left(-\frac{\partial \gamma_{xy}}{\partial z} + \frac{\partial \gamma_{xz}}{\partial y} + \frac{\partial \gamma_{yz}}{\partial x}\right) \tag{2.8.9c}$$

If one calls e_{ij} $(i,j = 1,2,3)$, the elements of the strain tensor, and x_i $(i = 1,2,3)$, the coordinate axes, these expressions can be written as

$$\frac{\partial^2 e_{ii}}{\partial x_j^2} + \frac{\partial^2 e_{jj}}{\partial x_i^2} = \frac{\partial^2 e_{ij}}{\partial x_i \partial x_j} \tag{2.8.10a}$$

$$2\frac{\partial^2 e_{ii}}{\partial x_j \partial x_k} = \frac{\partial}{\partial x_i}\left(\frac{\partial e_{ij}}{\partial x_k} + \frac{\partial e_{ik}}{\partial x_j} - \frac{\partial e_{jk}}{\partial x_i}\right) \tag{2.8.10b}$$

These are called the *compatibility conditions*, and they express the concept that the elements of the strain tensor **E** are not independent, but that the expressions that define them must fulfil the conditions that result from equations (2.8.9). Given that all of them contain second derivatives, it is evident that if the expressions that define the elements of **E** are linear or constant, they will always satisfy them.

EXERCISES

2.1. In the following figure, a cantilever beam appears whose points have experienced displacements, defined by the vector $\mathbf{r} = [r_1 r_2]^T$ whose expressions are indicated later.

1. Obtain the Green–Lagrange strain tensor.
2. Represent the changes in the strain components e_{11}, e_{12} and e_{22} according to δ at the points indicated in the figure.
3. Obtain the Cauchy strain tensor.
4. Explain the loads that are acting on the beam.

$$r_1 = x_1 - \frac{3\delta x_1 x_2}{(1+c)\,l^2} + \frac{3\delta x_1^2 x_2}{2(1+c)\,l^3}$$

$$r_2 = x_2 + \frac{3\delta x_1^2}{2(1+c)\,l^2} - \frac{\delta x_1^3}{2(1+c)\,l^3} + \frac{\delta c x_1}{(1+c)\,l}$$

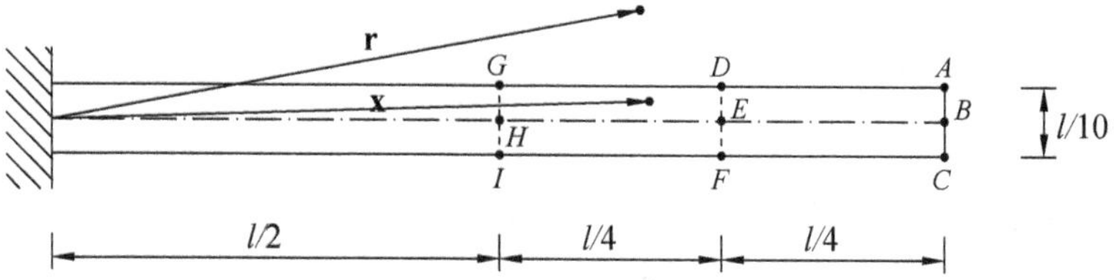

2.2. The points of a body experience some displacement defined by the functions

$$u = 4y^2 \qquad v = 2x^2 \qquad w = z^2$$

Find the following:

1. Expressions for the Cauchy strain tensor at any point.
2. Displacement vector at the point $P(1, 1, 1)$.
3. Principal directions of strain at the point P.
4. Graphical representation of Mohr's circles at point P.
5. Value of the maximum angular strain.
6. Normal strain of an elementary line element with origin in P in a direction that forms equal angles with the three coordinate axes.
7. Equations of strain quadrics.
8. Strain vector that corresponds to the direction of section 6 (use the indicatrix quadric).
9. Using the directrix quadric, find the direction that corresponds to a strain vector defined by

$$\boldsymbol{\varepsilon}_n = \left(6\sqrt{2}\mathbf{i} + 2\mathbf{j} + 0\mathbf{k}\right)/\sqrt{3}$$

2.3. Verify for what values of C, K and a the following state of strain is admissible as a solution to an elastic problem and obtain the field of displacements u, v and w.

$$\begin{aligned}
\varepsilon_x &= K\left(x^2 + y^2\right) + C\left(yz + z^2\right) & \gamma_{xy} &= aKxy + 2Cxz \\
\varepsilon_y &= Ky^2 + C\left(x + y + z\right) & \gamma_{xz} &= 2Cxz \\
\varepsilon_z &= C\left(y^2 + z^2\right) & \gamma_{yz} &= 2Cyz
\end{aligned}$$

2.4. If the strain is defined as λ, where $\lambda = l_r/l$, obtain, for the right tensor of Cauchy–Green $\mathbf{C}_r$ and for incompressible solids:

1. The value of the third invariant.
2. The expression of the invariants in states of uniaxial strain ε_x or of biaxial strain $\varepsilon_x = \varepsilon_y$.

2.5. The points of a solid elastic have experienced the following field of displacements.

$$u = 4zx \qquad v = -zx \qquad w = x^3$$

Find:

1. The Green–Lagrange strain tensor.
2. The Cauchy strain tensor.
3. The strains and principal directions of the Cauchy tensor at the point $(2, 0, 0)$.
4. Representation of Mohr's circles.
5. Expressions of the quadrics associated with the strains.
6. Components of the strain in the direction that forms equal angles with the coordinate axes of the first quadrant.

CHAPTER 3

EQUILIBRIUM EQUATIONS IN DEFORMABLE BODIES

3.1 STRESS TENSOR AT A POINT

Consider a continuum media that occupies a volume V in a three-dimensional space with Cartesian coordinates subjected to volume forces **b**, surface forces **p** and accelerations **a** whose values can be variable at each point. The expressions for these are

$$\mathbf{b} = \left[f_x(x,y,z,t), f_y(x,y,z,t), f_z(x,y,z,t)\right] \tag{3.1.1a}$$

$$\mathbf{p} = \left[p_x(x,y,z,t), p_y(x,y,z,t), p_z(x,y,z,t)\right] \tag{3.1.1b}$$

$$\mathbf{a} = \left[a_x(x,y,z,t), a_y(x,y,z,t), a_z(x,y,z,t)\right] \tag{3.1.1c}$$

Under the influence of these loads, each of the points of the body will change position, and the work carried out by the external loads will be equal to the intermolecular forces, which will be stored in the form of internal energy.

If one divides volume V with a cut in any π plane, defined by the vector **n**, two areas will be obtained, V_1, V_2, each of which will also be in equilibrium.

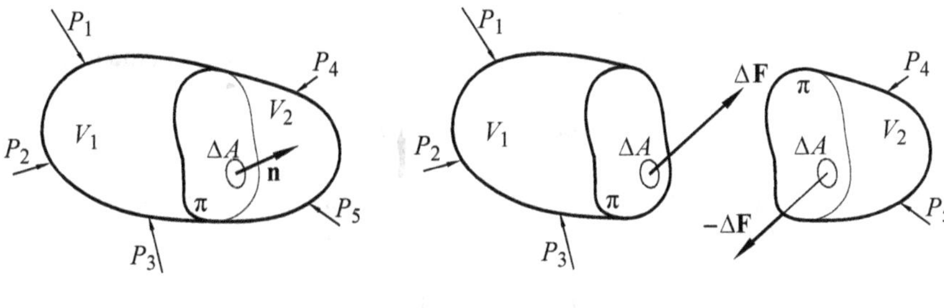

a) Deformable body and acting loads b) Equilibrium of two volumes of the body

Figure 3.1.1: Deformable body in equilibrium

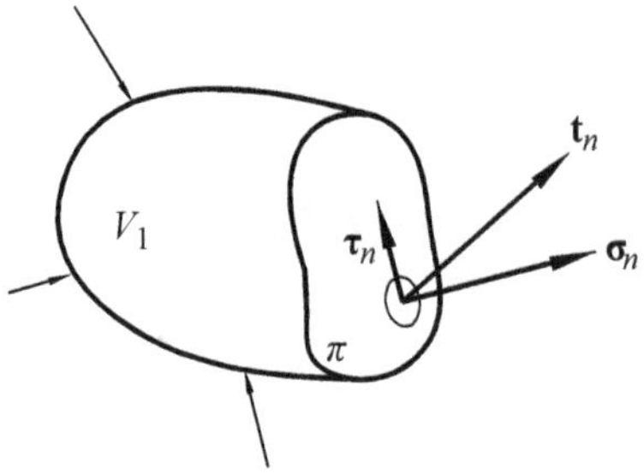

Figure 3.1.2: Normal and shear components of the stress vector

The equilibrium of V_1 is obtained when the existing loads in that volume and the interaction of V_2 are compensated. The equilibrium of the volume V_2 is obtained in a similar way. In each elementary surface ΔA of the common section, the interaction consists of two forces $\Delta\mathbf{F}$ and $-\Delta\mathbf{F}$. From the above, one can define the concept of *stress* at a point of the section by any π plane as

$$\mathbf{t}_n = \lim_{\Delta A \to 0} \frac{\Delta \mathbf{F}}{\Delta A} \tag{3.1.2}$$

because the elementary surface ΔA and the force $\Delta\mathbf{F}$ depend on the chosen direction $\mathbf{n}$, the resulting stress $\mathbf{t}_n$ will be different for each orientation of the cross-section.

Vector $\mathbf{t}_n$ can be decomposed into vector $\boldsymbol{\sigma}_n$, which is perpendicular to the plane, called the *normal stress*, and another $\boldsymbol{\tau}_n$ located on the plane of the cross-section, which is the *shear stress*.

$$\mathbf{t}_n = \boldsymbol{\sigma}_n + \boldsymbol{\tau}_n \tag{3.1.3}$$

It can also be decomposed according to the coordinate axes in the form

$$\mathbf{t}_n = t_{nx}\mathbf{i} + t_{ny}\mathbf{j} + t_{nz}\mathbf{k} \tag{3.1.4}$$

where t_{nx}, t_{ny}, t_{nz} are the projections of the vector $\mathbf{t}_n$ in the corresponding directions. Expressing equation (3.1.4) for these directions gives

$$\mathbf{t}_x = t_{xx}\mathbf{i} + t_{xy}\mathbf{j} + t_{xz}\mathbf{k} \tag{3.1.5a}$$
$$\mathbf{t}_y = t_{yx}\mathbf{i} + t_{yy}\mathbf{j} + t_{yz}\mathbf{k} \tag{3.1.5b}$$
$$\mathbf{t}_z = t_{zx}\mathbf{i} + t_{zy}\mathbf{j} + t_{zz}\mathbf{k} \tag{3.1.5c}$$

The stress vectors $\mathbf{t}_x, \mathbf{t}_y, \mathbf{t}_z$ can therefore be written as

$$\mathbf{t}_x = \begin{bmatrix} t_{xx} \\ t_{xy} \\ t_{xz} \end{bmatrix} \quad \mathbf{t}_y = \begin{bmatrix} t_{yx} \\ t_{yy} \\ t_{yz} \end{bmatrix} \quad \mathbf{t}_z = \begin{bmatrix} t_{zx} \\ t_{zy} \\ t_{zz} \end{bmatrix} \tag{3.1.6}$$

Defining a generic direction $\mathbf{n}$ by the cosine directors l, m, n of the angles that it forms with the coordinate axes, one can obtain the stress components $\mathbf{t}_n$ from the matrix $\mathcal{T}$

$$\mathcal{T} = \begin{bmatrix} t_{xx} & t_{yx} & t_{zx} \\ t_{xy} & t_{yy} & t_{zy} \\ t_{xz} & t_{yz} & t_{zz} \end{bmatrix} = \begin{bmatrix} \mathbf{t}_x & \mathbf{t}_y & \mathbf{t}_z \end{bmatrix} \tag{3.1.7}$$

and the following matrix relationship

$$\mathbf{t}_n = \mathcal{T}\mathbf{n} = \begin{bmatrix} t_{nx} \\ t_{ny} \\ t_{nz} \end{bmatrix} = \begin{bmatrix} t_{xx} & t_{yx} & t_{zx} \\ t_{xy} & t_{yy} & t_{yz} \\ t_{xz} & t_{yz} & t_{zz} \end{bmatrix} \begin{bmatrix} l \\ m \\ n \end{bmatrix} \tag{3.1.8}$$

In the case of the x axis, one has $l = 1$, $m = n = 0$, and, applying (3.1.8), one obtains equation (3.1.5a). The same verification can be carried out for the remaining coordinate axes.

3.2 INTERNAL EQUILIBRIUM EQUATIONS

Figure 3.2.1 shows a prism of dimensions dx, dy, dz of a material whose density is ρ. Equilibrium will be achieved between the forces existing on the boundary, the volume forces $\mathbf{b}$ and the accelerations $\mathbf{a}$, which are supposed to be constant in the differential volume. The stresses on the frontal element sides are considered positive when the directions coincide with the axis of the coordinate systems; however, on the other sides it is the other way around. The volume forces and the accelerations are considered positive when they coincide with the coordinate directions.

If equilibrium of the forces is established in the direction x, the following equation is obtained

$$\begin{aligned} &\left(t_{xx} + \frac{\partial t_{xx}}{\partial x}dx - t_{xx}\right)dydz + \left(t_{yx} + \frac{\partial t_{yx}}{\partial y}dy - t_{yx}\right)dxdz \\ &+ \left(t_{zx} + \frac{\partial t_{zx}}{\partial z}dz - t_{zx}\right)dxdy + b_x dxdydz - \rho a_x dxdydz = 0 \end{aligned} \tag{3.2.1}$$

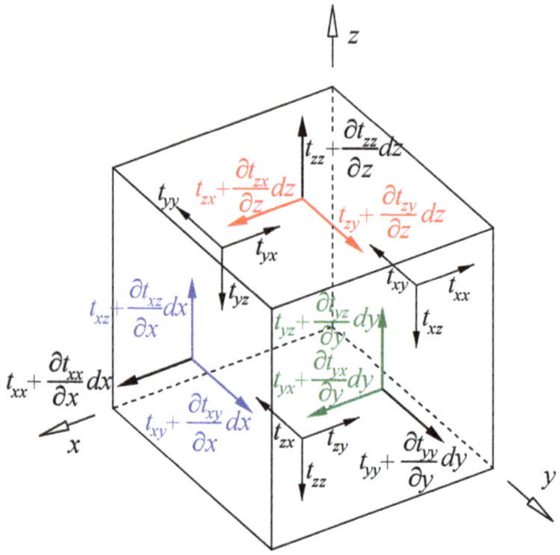

Figure 3.2.1: Equilibrium of forces in a differential volume

simplifying terms

$$\frac{\partial t_{xx}}{\partial x}+\frac{\partial t_{yx}}{\partial y}+\frac{\partial t_{zx}}{\partial z}+b_x-\rho a_x=0 \tag{3.2.2a}$$

establishing the equilibrium with respect to the other axes, similar equations are obtained

$$\frac{\partial t_{xy}}{\partial x}+\frac{\partial t_{yy}}{\partial y}+\frac{\partial t_{zy}}{\partial z}+b_y-\rho a_y=0 \tag{3.2.2b}$$

$$\frac{\partial t_{xz}}{\partial x}+\frac{\partial t_{yz}}{\partial y}+\frac{\partial t_{zz}}{\partial z}+b_z-\rho a_z=0 \tag{3.2.2c}$$

These equations are called *internal equilibrium equations.* If $t_{ij}\,(i,j=1,2,3)$ are defined as the elements of the stress tensor $\boldsymbol{\tau}$, the coordinate axes are called $x_i\,(i=1,2,3)$, and $b_i, a_i\,(i=1,2,3)$ are the components of the volume and acceleration forces, the above equations are expressed as

$$\frac{\partial t_{ij}}{\partial x_i}+b_j-\rho a_j=0 \quad i,j=1,2,3 \tag{3.2.3}$$

or, using the *divergence* operator

$$\nabla\boldsymbol{\tau}+\mathbf{b}-\rho\mathbf{a}=\mathbf{0} \tag{3.2.4}$$

The equilibrium of moments in the direction x at the centre of gravity leads to

$$t_{yz}dxdz\frac{dy}{2}+\left(t_{yz}+\frac{\partial t_{yz}}{\partial y}dy\right)dxdz\frac{dy}{2}$$
$$-t_{zy}dxdy\frac{dz}{2}-\left(t_{zy}+\frac{\partial t_{zy}}{\partial z}dz\right)dxdy\frac{dz}{2}=0 \qquad (3.2.5)$$

disregarding the differentials of higher order, one obtains

$$t_{yz}=t_{zy} \qquad (3.2.6a)$$

and, similarly, it turns out that

$$t_{zx}=t_{xz} \qquad (3.2.6b)$$
$$t_{xy}=t_{yx} \qquad (3.2.6c)$$

The expressions (3.2.6) result in the matrix (3.1.7) being symmetric. The elements of the main diagonal are normal stresses and the remaining elements are shear stresses. Using the usual terminology in engineering, equation (3.1.8) becomes

$$\begin{bmatrix} t_{nx} \\ t_{ny} \\ t_{nz} \end{bmatrix}=\begin{bmatrix} \sigma_x & \tau_{xy} & \tau_{xz} \\ \tau_{xy} & \sigma_y & \tau_{yz} \\ \tau_{xz} & \tau_{yz} & \sigma_z \end{bmatrix}\begin{bmatrix} l \\ m \\ n \end{bmatrix} \qquad (3.2.7)$$

or, in matrix form

$$\mathbf{t}_n=\boldsymbol{\tau}\mathbf{n} \qquad (3.2.8)$$

In figure 3.2.2, the stresses on a differential volume are shown using the terminology that has been adopted.

One can demonstrate that $\boldsymbol{\tau}$ is a tensor of second order, which is called the *Cauchy stress tensor*. From (3.2.8), the normal component of the stress $\mathbf{t}_n$ is obtained as

$$\sigma=\mathbf{n}^T\mathbf{t}_n=\mathbf{n}^T\boldsymbol{\tau}\mathbf{n} \qquad (3.2.9)$$

developing (3.2.9), one achieves

$$\sigma=\sigma_x l^2+\sigma_y m^2+\sigma_z n^2+2\tau_{xy}lm+2\tau_{xz}ln+2\tau_{yz}mn \qquad (3.2.10)$$

The modulus τ of the shear stress is

$$\tau=\sqrt{|\mathbf{t}_n|^2-\sigma^2} \qquad (3.2.11)$$

and is defined by the vector

$$\boldsymbol{\tau}_n = \mathbf{t}_n - \sigma \mathbf{n} \tag{3.2.12}$$

Let us assume that the current coordinate system is substituted by another whose transformation matrix is $\mathbf{T}$, similarly to how it was shown when studying body deformations in equation (2.3.34).

The direction vector $\mathbf{n}$ and the stress vector $\mathbf{t}_n$ in the new axes will be

$$\bar{\mathbf{n}} = \mathbf{T}\mathbf{n} \quad \bar{\mathbf{t}}_n = \mathbf{T}\mathbf{t}_n \tag{3.2.13}$$

and the stress tensor in the new coordinate system will be

$$\bar{\boldsymbol{\tau}} = \mathbf{T}\boldsymbol{\tau}\mathbf{T}^T \tag{3.2.14}$$

Other definitions of the stress tensor exist that are useful for some types of materials. If the determinant of the matrix of the position vector gradients $\mathbf{J}$ defined in (2.1.10) is denoted by J

$$J = |\mathbf{J}| \tag{3.2.15}$$

the *Kirchhoff stress tensor* can be defined as

$$\boldsymbol{\tau}_k = J\boldsymbol{\tau} \tag{3.2.16}$$

This tensor changes the modulus of the elements of the Cauchy stress tensor. As $\mathbf{J} = \mathbf{I} + \mathbf{J}_d$, in the case of small strains the value of $J \approx 1$.

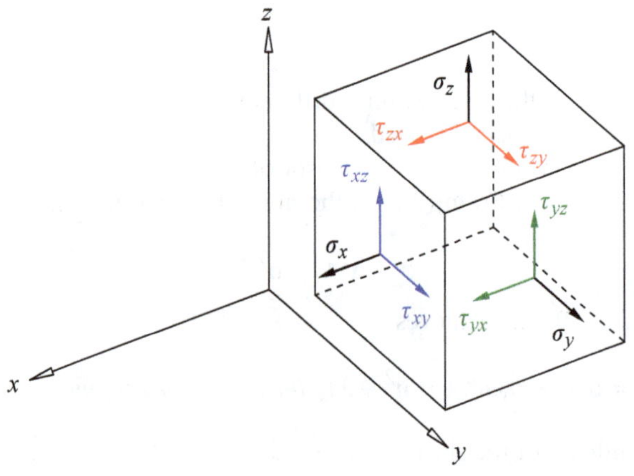

Figure 3.2.2: Stresses in the directions of the coordinate planes

Also of interest is the *first Piola–Kirchhoff stress tensor*, defined as

$$\boldsymbol{\tau}_{P1} = J\boldsymbol{\tau}\mathbf{J}^{-1^T} \tag{3.2.17}$$

in this case, one can verify that the tensor $\boldsymbol{\tau}_{P1}$ is not symmetrical.

Similarly, one can define the *second Piola–Kirchhoff tensor* $\boldsymbol{\tau}_{P2}$, which is symmetric, i.e.

$$\boldsymbol{\tau}_{P2} = J\mathbf{J}^{-1}\boldsymbol{\tau}\mathbf{J}^{-1^T} \tag{3.2.18}$$

and therefore

$$\boldsymbol{\tau} = \frac{1}{J}\mathbf{J}\boldsymbol{\tau}_{P2}\mathbf{J}^T \tag{3.2.19}$$

taking into consideration (3.2.16), the Kirchhoff stress tensor can be expressed as

$$\boldsymbol{\tau}_k = \mathbf{J}\boldsymbol{\tau}_{P2}\mathbf{J}^T \tag{3.2.20}$$

3.3 EQUATIONS OF EQUILIBRIUM ON THE BOUNDARY

Let us consider a differential volume at a point on the boundary, defined by a tetrahedron whose sides are the tangent plane to the continuum at that point and the three coordinate planes. In this volume, the external forces **p** that act on the surface, the volume forces **b**, the accelerations **a** and the forces on the coordinate planes, should be balanced.

The area on the tangent plane is dA. One assumes that the boundary is smooth enough and that there are no discontinuities in the first two orders of differentiation. The equilibrium equations in each axis are

$$\frac{\sigma_x}{2}dydz + \frac{\tau_{xy}}{2}dxdz + \frac{\tau_{xz}}{2}dxdy - p_x dA - (b_x - \rho a_x)\frac{dxdydz}{6} = 0 \tag{3.3.1a}$$

$$\frac{\tau_{xy}}{2}dydz + \frac{\sigma_y}{2}dxdz + \frac{\tau_{xz}}{2}dxdy - p_y dA - (b_y - \rho a_y)\frac{dxdydz}{6} = 0 \tag{3.3.1b}$$

$$\frac{\tau_{xz}}{2}dydz + \frac{\tau_{yz}}{2}dxdz + \frac{\sigma_z}{2}dxdy - p_z dA - (b_z - \rho a_z)\frac{dxdydz}{6} = 0 \tag{3.3.1c}$$

it is known that

$$\frac{1}{2}dydz = ldA \quad \frac{1}{2}dxdz = mdA \quad \frac{1}{2}dxdy = ndA \tag{3.3.2}$$

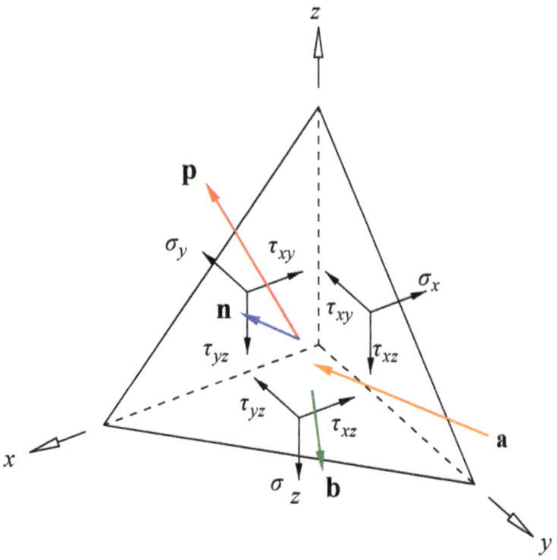

Figure 3.3.1: Equilibrium of forces on the boundary

if one ignores the differentials of higher order, the volume forces **b** and the accelerations **a**, expressions (3.3.1) and (3.3.2) can be combined as

$$\begin{bmatrix} p_x \\ p_y \\ p_z \end{bmatrix} = \begin{bmatrix} \sigma_x & \tau_{xy} & \tau_{xz} \\ \tau_{xy} & \sigma_y & \tau_{yz} \\ \tau_{xz} & \tau_{yz} & \sigma_z \end{bmatrix} \begin{bmatrix} l \\ m \\ n \end{bmatrix} \tag{3.3.3}$$

and, in matrix notation

$$\mathbf{p} = \boldsymbol{\tau}\mathbf{n} \tag{3.3.4}$$

Comparing (3.3.4) and (3.2.8), one observes that at the points of the boundary the forces of the surface coincide with the stress on those points. Equation (3.3.4) is called the *boundary equilibrium equation.*

3.4 PRINCIPAL STRESSES AND PRINCIPAL STRESS DIRECTIONS

It is always possible to find a system of coordinate axes in which only the normal component of the stresses exists in the three coordinate planes, which will convert the matrix $\boldsymbol{\tau}$ of the stress tensor into a diagonal matrix. To find the directions that those new axes will form with the current trihedron of coordinates, one simply has to consider that, when studying these directions, the resulting stress will be parallel to the normal vector **n**. Therefore

$$\mathbf{t}_n = \sigma\mathbf{n} = \boldsymbol{\tau}\mathbf{n} \tag{3.4.1}$$

developing (3.4.1)

$$\sigma l = \sigma_x l + \tau_{xy} m + \tau_{xz} n \tag{3.4.2a}$$

$$\sigma m = \tau_{xy} l + \sigma_y m + \tau_{yz} n \tag{3.4.2b}$$

$$\sigma n = \tau_{xz} l + \tau_{yz} m + \sigma_z n \tag{3.4.2c}$$

The linear system of equations (3.4.2) is homogeneous and, for a solution to exist, the following condition must be met

$$(\boldsymbol{\tau} - \sigma \mathbf{I})\,\mathbf{n} = \mathbf{0} \tag{3.4.3a}$$

or, similarly

$$\begin{vmatrix} \sigma_x - \sigma & \tau_{xy} & \tau_{xz} \\ \tau_{xy} & \sigma_y - \sigma & \tau_{yz} \\ \tau_{xz} & \tau_{yz} & \sigma_z - \sigma \end{vmatrix} = 0 \tag{3.4.3b}$$

The determinant of (3.4.3) provides an equation of third degree, which is usually written in the form

$$\sigma^3 - J_1 \sigma^2 + J_2 \sigma - J_3 = 0 \tag{3.4.4}$$

The values of the coefficients J_1, J_2, J_3 are independent of the coordinate axes. They are called the *invariants of the stress tensor* and their expressions are

linear invariant $\quad J_1 = \sigma_x + \sigma_y + \sigma_z \quad$ (3.4.5a)

quadratic invariant $\quad J_2 = \sigma_x \sigma_y + \sigma_x \sigma_z + \sigma_y \sigma_z - \tau_{xy}^2 - \tau_{xz}^2 - \tau_{yz}^2 \quad$ (3.4.5b)

cubic invariant $\quad J_3 = |\boldsymbol{\tau}| \quad$ (3.4.5c)

The three roots σ_1, σ_2, σ_3 of (3.4.4) are necessarily real due to its being the characteristic equation of a symmetrical matrix, and they are denoted as the *principal stresses.* The directions associated with each of them are called the *principal directions* and are obtained by replacing each value σ_i $(i = 1, 2, 3)$ in the system (3.4.2), complemented with the equation

$$l^2 + m^2 + n^2 = 1 \tag{3.4.6}$$

In two-dimensional stress fields, the dimension of the tensor $\boldsymbol{\tau}$ is reduced. For example, when only σ_x, τ_{xy}, σ_y stresses exist, the stress tensor is

$$\boldsymbol{\tau} = \begin{bmatrix} \sigma_x & \tau_{xy} \\ \tau_{xy} & \sigma_y \end{bmatrix} \tag{3.4.7}$$

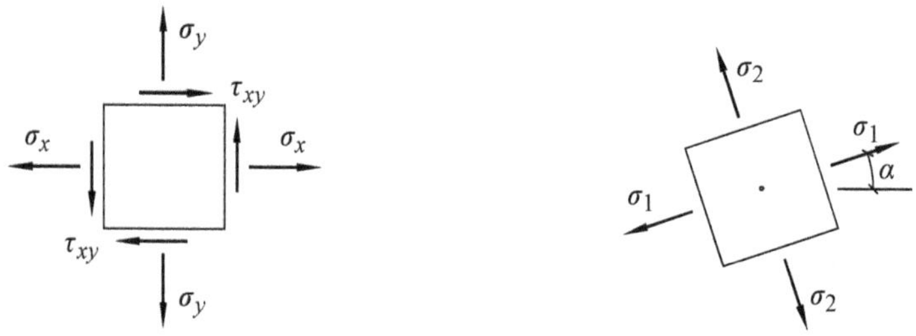

a) Stresses according to the coordinate axes b) Principal directions and stresses

Figure 3.4.1: Two-dimensional stress field

If the principal stresses are calculated in the way shown above, one obtains the following results

$$\sigma_1 = \frac{\sigma_x + \sigma_y}{2} + \sqrt{\left(\frac{\sigma_x - \sigma_y}{2}\right)^2 + \tau_{xy}^2} \quad \sigma_2 = \frac{\sigma_x + \sigma_y}{2} - \sqrt{\left(\frac{\sigma_x - \sigma_y}{2}\right)^2 + \tau_{xy}^2} \tag{3.4.8}$$

the direction that the vector σ_1 forms with the x axis is defined by the equation

$$\tan(2\alpha) = \frac{2\tau_{xy}}{\sigma_x - \sigma_y} \tag{3.4.9}$$

3.5 MAXIMUM SHEAR STRESSES

It is also interesting to know the planes in which the shear stress will have its maximum value. The coordinate system that coincides with the principal directions will be used to identify them, in order to simplify the expressions. In those axes, the stress corresponding to a plane defined by the direction **n** and the normal and shear components of the stress are

$$\mathbf{t}_n = \mathfrak{T}\mathbf{n} = \sigma_1 l\mathbf{i} + \sigma_2 m\mathbf{j} + \sigma_3 n\mathbf{k} \tag{3.5.1}$$

$$\sigma = \mathbf{n}^T \mathbf{t}_n = \sigma_1 l^2 + \sigma_2 m^2 + \sigma_3 n^2 \tag{3.5.2}$$

$$\tau^2 = |\mathbf{t}_n|^2 - \sigma^2 = \left(\sigma_1^2 l^2 + \sigma_2^2 m^2 + \sigma_3^2 n^2\right) - \left(\sigma_1 l^2 + \sigma_2 m^2 + \sigma_3 n^2\right)^2 \tag{3.5.3}$$

The values of l, m, n, which maximize equation (3.5.3), are obtained by defining a Lagrangian function in the form

$$L = \sigma_1^2 l^2 + \sigma_2^2 m^2 + \sigma_3^2 n^2 - \left(\sigma_1 l^2 + \sigma_2 m^2 + \sigma_3 n^2\right)^2 + \lambda\left(l^2 + m^2 + n^2 - 1\right) \tag{3.5.4}$$

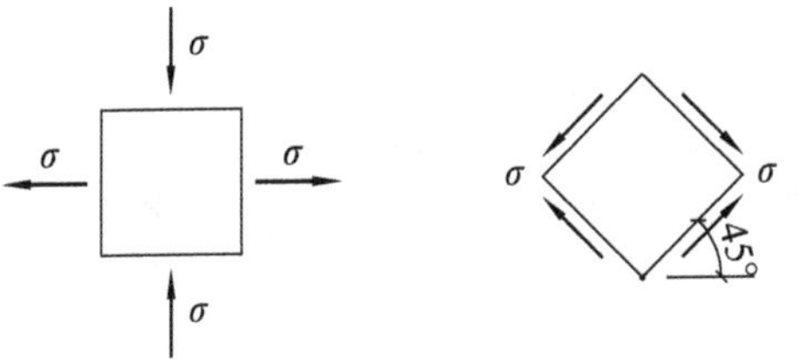

a) Principal stresses b) Maximum shear stress

Figure 3.5.1: Directions of the principal stresses and the maximum shear stress

the maximum values of (3.5.4) shall comply with the conditions

$$\frac{\partial L}{\partial l} = \frac{\partial L}{\partial m} = \frac{\partial L}{\partial n} = \frac{\partial L}{\partial \lambda} = 0 \quad (3.5.5)$$

One can verify that the solution that maximizes (3.5.4) corresponds to the values

$$m = 0 \quad (3.5.6a)$$

$$l^2 = n^2 = 0.5 \quad (3.5.6b)$$

The resulting planes of expression (3.5.6) form angles of 45° with those of the principal directions σ_1, σ_3 in the σ_2 direction.

The value of the maximum shear stress is

$$\tau_{max} = \frac{\sigma_1 - \sigma_3}{2} \quad (3.5.7)$$

A particular case of interest occurs when the values of the principal stresses are

$$\sigma_1 = \sigma \quad \sigma_2 = 0 \quad \sigma_3 = -\sigma \quad (3.5.8)$$

This situation corresponds to a stress field of traction in a direction and a stress field of compression of equal value in the other principal direction. The value of the maximum shear stress is

$$\tau_{max} = \sigma \quad (3.5.9)$$

therefore, the modulus of the maximum shear stress in that stress field coincides with the modulus of the two principal stresses that are not null, and in that direction the normal stress is zero.

3.6 QUADRICS ASSOCIATED WITH THE STRESS TENSOR

For the stress tensor, one can establish the same set of surfaces as for the strain tensor described in chapter 2. All of them have the same properties as those previously identified.

3.6.1 Ellipsoid of stress

This is the locus of the ends of the total stress vectors $\mathbf{t}_n$ at a point that corresponds to the directions $\mathbf{n}$.

One can obtain it by recalling equation (3.2.7). In the case that the coordinate axes correspond to the principal directions, one has

$$\begin{bmatrix} t_{nx} \\ t_{ny} \\ t_{nz} \end{bmatrix} = \begin{bmatrix} \sigma_1 & 0 & 0 \\ 0 & \sigma_2 & 0 \\ 0 & 0 & \sigma_3 \end{bmatrix} \begin{bmatrix} l \\ m \\ n \end{bmatrix} \tag{3.6.1.1}$$

$$l^2 + m^2 + n^2 = 1 \tag{3.6.1.2}$$

eliminating l, m, n from those equations gives

$$\frac{t_{nx}^2}{\sigma_1^2} + \frac{t_{ny}^2}{\sigma_2^2} + \frac{t_{nz}^2}{\sigma_3^2} = 1 \tag{3.6.1.3}$$

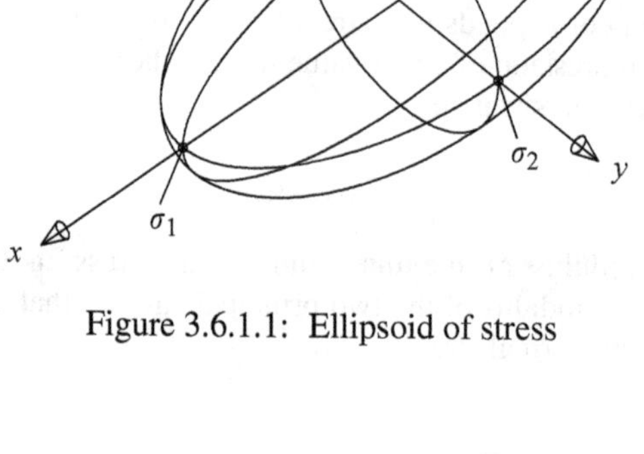

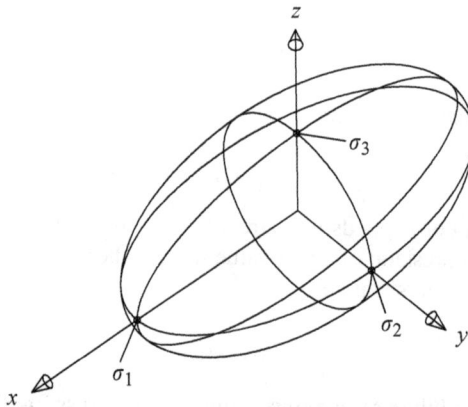

Figure 3.6.1.1: Ellipsoid of stress

which is usually written as

$$\frac{x^2}{\sigma_1^2} + \frac{y^2}{\sigma_2^2} + \frac{z^2}{\sigma_3^2} = 1 \qquad (3.6.1.4)$$

3.6.2 Indicatrix quadric of stresses

This is the locus of the ends of the segments ON along each direction $\mathbf{n}$ measured from the origin of coordinates.

$$ON = \frac{1}{\sqrt{|\sigma|}} \qquad (3.6.2.1)$$

The coordinates of the point N are

$$x = \frac{l}{\sqrt{|\sigma|}} \quad y = \frac{m}{\sqrt{|\sigma|}} \quad z = \frac{n}{\sqrt{|\sigma|}} \qquad (3.6.2.2)$$

if the coordinate axes are the principal directions, it turns out that

$$\sigma = \sigma_1 l^2 + \sigma_2 m^2 + \sigma_3 n^2 \qquad (3.6.2.3)$$

eliminating l, m, n in these equations gives

$$\sigma_1 x^2 + \sigma_2 y^2 + \sigma_3 z^2 = \frac{\sigma}{|\sigma|} = c = \pm 1 \qquad (3.6.2.4)$$

The surface that represents this equation is called the *indicatrix quadric*, and the same discussion as in section 2.5 of chapter 2 follows.

a) If $\sigma_i > 0$ $(i = 1,2,3)$ for $c = -1$, equation (3.6.2.4) is an imaginary ellipsoid that has no physical meaning. For $c = 1$, one obtains a real ellipsoid that is the locus of the points ON. All the normal stresses will be of traction, and the stress vector $\mathbf{t}_n$ forms an acute angle with the direction $\mathbf{n}$.
b) If $\sigma_i < 0$ $(i = 1,2,3)$, the real ellipsoid appears for $c = -1$. All the normal stresses are in compression, and the stress vector $\mathbf{t}_n$ forms an obtuse angle with direction $\mathbf{n}$.
c) If the two principal stresses are positive, $\sigma_1 \geq \sigma_2 > 0$ and $\sigma_3 < 0$, for $c = 1$, a hyperboloid of one sheet is the result, and for $c = -1$, one obtains a hyperboloid of two sheets, as shown in figure 3.6.2.2.
   If a direction $\mathbf{n}$ cuts the hyperboloid of one sheet, the normal stress is of traction and when it cuts that of two sheets it is of compression.

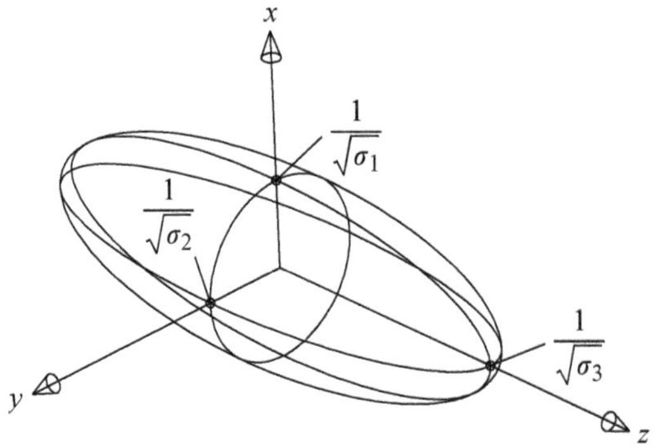

Figure 3.6.2.1: Indicatrix ellipsoid of stresses

Both hyperboloids are separated by an asymptotic cone, which has the equation

$$\sigma_1 x^2 + \sigma_2 y^2 + \sigma_3 z^2 = 0 \qquad (3.6.2.5)$$

If the direction **n** coincides with one of the generatrices of the cone, the normal stress shall be equal to zero.

d) If a principal stress is positive, $\sigma_1 > 0$, and the other two are negative, $\sigma_3 \leq \sigma_2 < 0$, for $c = -1$, one obtains a hyperboloid of one sheet, and for $c = 1$, another of two sheets, and the conclusions are very similar to those of the previous case.

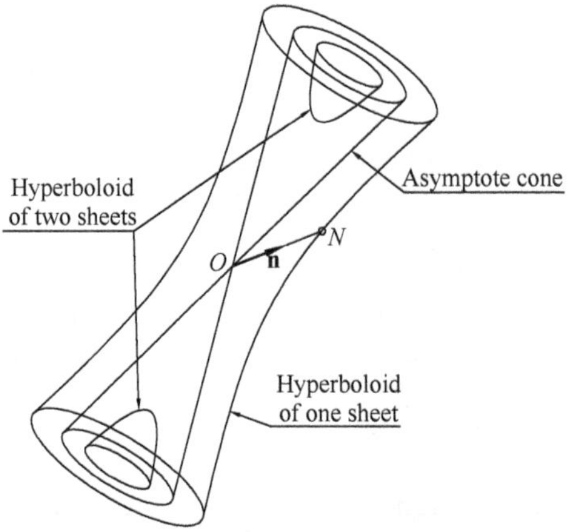

Figure 3.6.2.2: Indicatrix hyperboloid of stresses

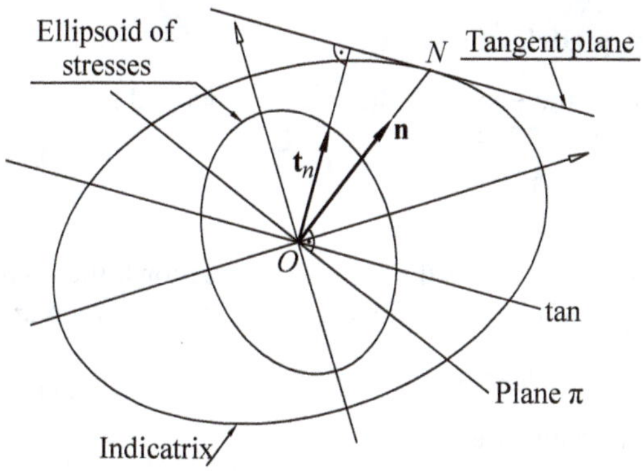

Figure 3.6.2.3: Direction **n** and stress $\mathbf{t}_n$ in the indicatrix ellipsoid of stress

The indicatrix quadric enables one to determine the stress that corresponds to a direction **n**. Figures 3.6.2.3 and 3.6.2.4 indicate how to obtain $\mathbf{t}_n$ from **n** in the ellipsoid and hyperboloid of stresses.

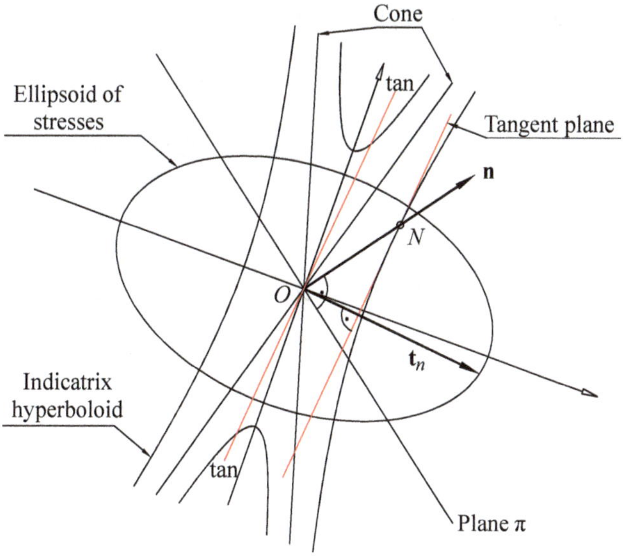

Figure 3.6.2.4: Direction **n** and stress $\mathbf{t}_n$ in the indicatrix hyperboloid of stress

3.6.3 Directrix quadric of stresses

This is the locus of the end N of the vectors obtained by modifying the stress vector $\mathbf{t}_n$ in the form

$$\frac{\mathbf{t}_n}{\sqrt{|\sigma|}} \tag{3.6.3.1}$$

If the coordinate axes are the principal directions, the components of the stress vector corresponding to a plane defined by direction $\mathbf{n}$ are

$$t_{nx} = \sigma_1 l \quad t_{ny} = \sigma_2 m \quad t_{nz} = \sigma_3 n \tag{3.6.3.2}$$

the coordinates of point N are

$$x = \frac{\sigma_1 l}{\sqrt{|\sigma|}} \quad y = \frac{\sigma_2 m}{\sqrt{|\sigma|}} \quad z = \frac{\sigma_3 n}{\sqrt{|\sigma|}} \tag{3.6.3.3}$$

and, taking into account equation (3.5.2), gives

$$\frac{x^2}{\sigma_1} + \frac{y^2}{\sigma_2} + \frac{z^2}{\sigma_3} = \pm 1 \tag{3.6.3.4}$$

This quadric has the attribute that if the direction of a stress vector $\mathbf{t}_n$, which provides point N of figure 3.6.3.2 in the directrix quadric, is known, it can be proven that it corresponds to the stress in the plane defined by the direction $\mathbf{n}$ which is normal to the tangent plane at point $\mathbf{t}_n$.

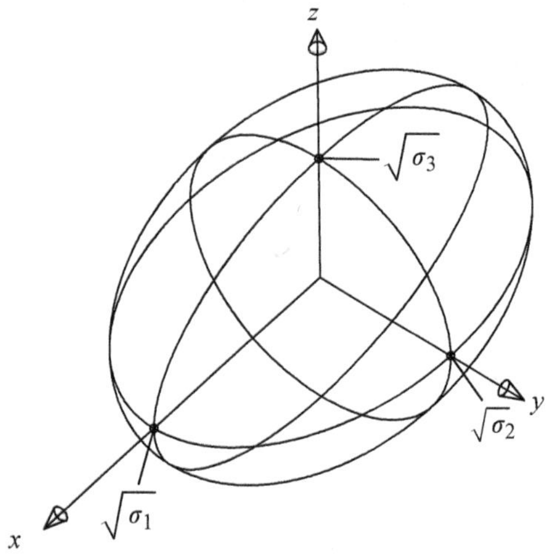

Figure 3.6.3.1: Directrix ellipsoid of stresses

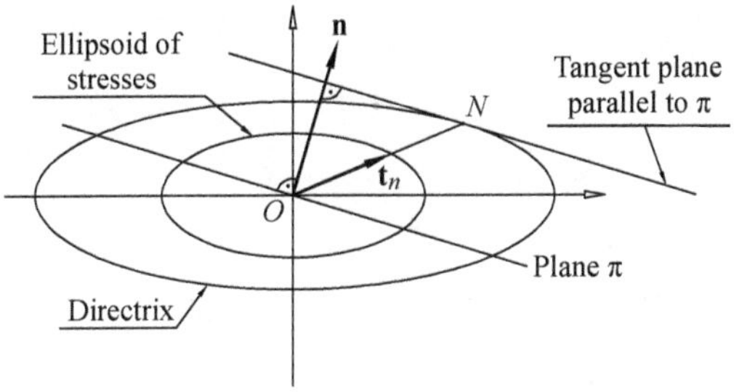

Figure 3.6.3.2: Plane section of the directrix ellipsoid of stresses

3.7 MOHR'S CIRCLES OF STRESS

If the coordinate axes are the principal directions, recalling (3.5.1) and (3.5.3) gives

$$|\mathbf{t}_n|^2 = (\sigma_1 l)^2 + (\sigma_2 m)^2 + (\sigma_3 n)^2 = \sigma^2 + \tau^2 \tag{3.7.1}$$

also, according to (3.5.2)

$$\sigma = \sigma_1 l^2 + \sigma_2 m^2 + \sigma_3 n^2 \tag{3.7.2}$$

if m and n are removed from these equations, and one divides by $(\sigma_2 - \sigma_3)$, recalling that $\sigma_1 \geq \sigma_2 \geq \sigma_3$, one finally obtains

$$\tau^2 + \left(\sigma - \frac{\sigma_2 + \sigma_3}{2}\right)^2 - \left(\frac{\sigma_2 - \sigma_3}{2}\right)^2 = l^2 (\sigma_1 - \sigma_2)(\sigma_1 - \sigma_3) \geq 0 \tag{3.7.3a}$$

if removing l and n and proceeding similarly, it is possible obtain

$$\tau^2 + \left(\sigma - \frac{\sigma_1 + \sigma_3}{2}\right)^2 - \left(\frac{\sigma_1 - \sigma_3}{2}\right)^2 = -m^2 (\sigma_2 - \sigma_3)(\sigma_1 - \sigma_2) \leq 0 \tag{3.7.3b}$$

finally, removing l and m in a similar way, the following expression is obtained

$$\tau^2 + \left(\sigma - \frac{\sigma_1 + \sigma_2}{2}\right)^2 - \left(\frac{\sigma_1 - \sigma_2}{2}\right)^2 = n^2 (\sigma_3 - \sigma_1)(\sigma_3 - \sigma_2) \geq 0 \tag{3.7.3c}$$

These equations can be represented by the circles and the grey area shown in figure 3.7.1, which is the locus of the ends of the stress vectors, defined by its components (σ, τ), giving rise to Mohr's circles of stress.

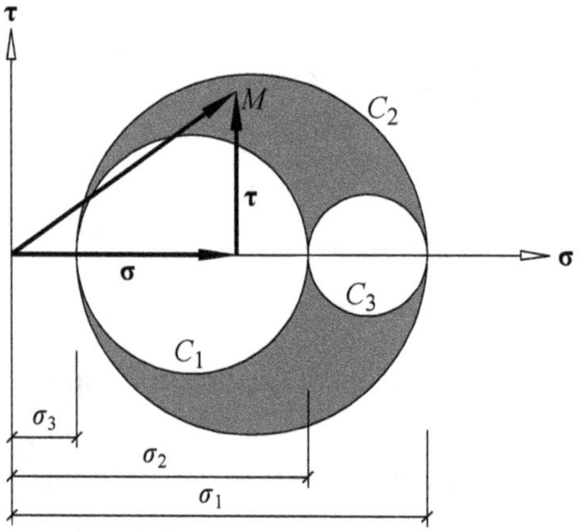

Figure 3.7.1: Mohr's circles of stress

Mohr's circles identify the point that corresponds to the stress vector of any plane defined by direction **n** that can be represented by its components with respect to the principal directions of stress $\mathbf{n} = l\mathbf{i} + m\mathbf{j} + n\mathbf{k}$. To this end, one can proceed as in chapter 2. If one defines

$$l = \cos\alpha \quad n = \cos\theta \tag{3.7.4}$$

and the angles α and θ are taken from the σ axis, as indicated in figure 3.7.2, the points P and Q are obtained. Tracing from the σ axis the circles with radius AP and BQ, respectively, one obtains the point M, which represents the stress vector for that direction, A and B being the centres of the inner circles.

This graphical construction also allows one to verify that the maximum shear stresses correspond to points M_1 and M_2, which are obtained with $\alpha = \theta = 45°$ as shown in figure 3.7.3. This means that

$$l = \cos\alpha = \pm\frac{\sqrt{2}}{2} \quad n = \cos\theta = \pm\frac{\sqrt{2}}{2} \tag{3.7.5}$$

and therefore $m = 0$, which is what was obtained analytically above. One can also verify that the maximum value is the one obtained before, i.e.

$$\tau_{max} = \frac{\sigma_1 - \sigma_3}{2} \tag{3.7.6}$$

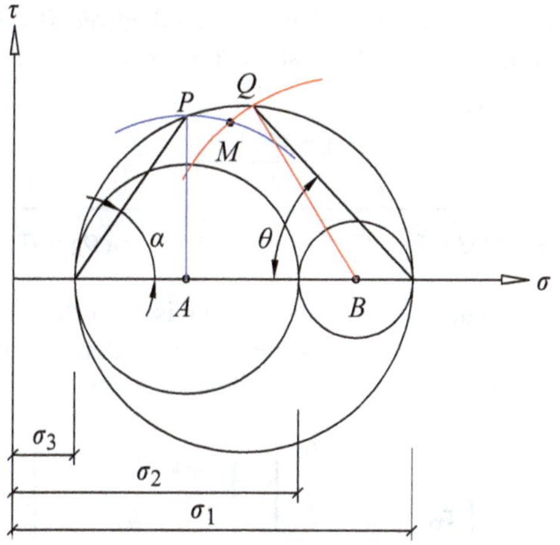

Figure 3.7.2: Graphical procedure to obtain the location of the stress vector

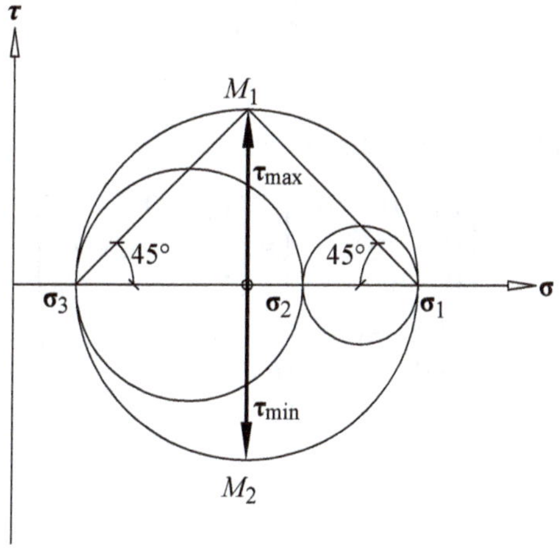

Figure 3.7.3: Values of the maximum shear stresses

3.8 SPHERICAL AND DEVIATORIC TENSORS

Suppose that a stress field exists, defined by the tensor $\boldsymbol{\tau}$ whose principal stresses are σ_i $(i = 1,2,3)$. Also consider the planes whose directions are $\mathbf{n}_{oct} = \left[\pm 1/\sqrt{3} \pm 1/\sqrt{3} \pm 1/\sqrt{3}\right]^T$, which constitute eight planes whose directions form equal angles with respect to the three coordinate axes, when these are the principal directions, and therefore form an octahedron. If one calculates the normal and shear components of the stress tensor in each of these planes, the result is

$$\sigma_{oct} = \frac{\sigma_1 + \sigma_2 + \sigma_3}{3} = \sigma_m \tag{3.8.1}$$

$$\tau_{oct} = \frac{1}{3}\sqrt{(\sigma_1 - \sigma_2)^2 + (\sigma_1 - \sigma_3)^2 + (\sigma_2 - \sigma_3)^2} \tag{3.8.2}$$

as one can see, the *octahedral normal stress* is also the mean value of the principal stresses, and for this reason they are also denoted as σ_m. From this, the stress tensor can be decomposed as follows

$$\boldsymbol{\tau} = \begin{bmatrix} \sigma_x & \tau_{xy} & \tau_{xz} \\ \tau_{xy} & \sigma_y & \tau_{yz} \\ \tau_{xz} & \tau_{yz} & \sigma_z \end{bmatrix} = \begin{bmatrix} \sigma_m & 0 & 0 \\ 0 & \sigma_m & 0 \\ 0 & 0 & \sigma_m \end{bmatrix} + \begin{bmatrix} \sigma_x - \sigma_m & \tau_{xy} & \tau_{xz} \\ \tau_{xy} & \sigma_y - \sigma_m & \tau_{yz} \\ \tau_{xz} & \tau_{yz} & \sigma_z - \sigma_m \end{bmatrix} \tag{3.8.3a}$$

or, in the principal directions of stress

$$\boldsymbol{\tau} = \begin{bmatrix} \sigma_1 & 0 & 0 \\ 0 & \sigma_2 & 0 \\ 0 & 0 & \sigma_3 \end{bmatrix} = \begin{bmatrix} \sigma_m & 0 & 0 \\ 0 & \sigma_m & 0 \\ 0 & 0 & \sigma_m \end{bmatrix} + \begin{bmatrix} \sigma_1 - \sigma_m & 0 & 0 \\ 0 & \sigma_2 - \sigma_m & 0 \\ 0 & 0 & \sigma_3 - \sigma_m \end{bmatrix} \tag{3.8.3b}$$

or simply written as

$$\boldsymbol{\tau} = \boldsymbol{\tau}_m + \boldsymbol{\tau}_d \tag{3.8.3c}$$

The stress field defined by $\boldsymbol{\tau}_m$ is commonly known as the *spherical stress field* because it has the same stress value in any direction. This is what happens to a solid immersed in water; thus, it is also known as the *hydrostatic stress field*. The tensor $\boldsymbol{\tau}_d$ is known as the *deviatoric stress field* because it is the difference between the stress field $\boldsymbol{\tau}$ and the spherical tensor $\boldsymbol{\tau}_m$.

The stress $\mathbf{t}_n$ that corresponds to generic direction $\mathbf{n}$ can be decomposed into the component $\boldsymbol{\sigma}_{n,m}$, produced by the spherical tensor, and $\mathbf{t}_{n,d}$ that corresponds to the deviatoric tensor.

$$\mathbf{t}_n = \boldsymbol{\sigma}_{n,m} + \mathbf{t}_{n,d} = \sigma_m \mathbf{n} + \mathbf{t}_{n,d} = \sigma_{oct}\mathbf{n} + \mathbf{t}_{n,d} \tag{3.8.4}$$

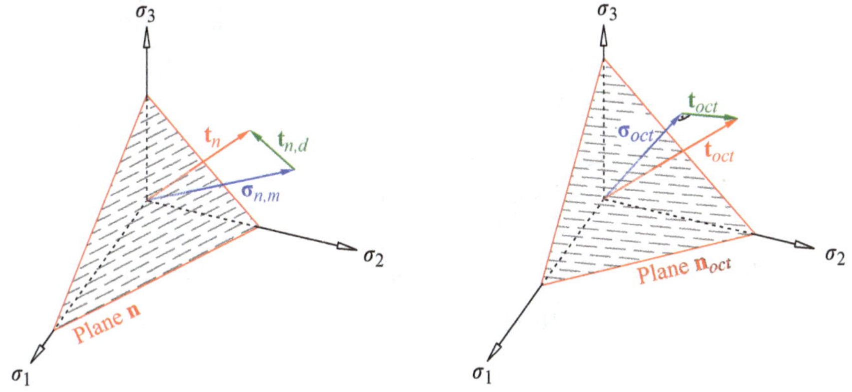

a) Stress according to direction **n** b) Stress according to direction $\mathbf{n}_{oct}$

Figure 3.8.1: Decomposition of stress $\mathbf{t}_n$ and $\mathbf{t}_{oct}$

If direction **n** is that of the $\mathbf{n}_{\mathrm{oct}}$ planes mentioned above, the expression that results is

$$\mathbf{t}_{oct} = \boldsymbol{\sigma}_{oct,m} + \mathbf{t}_{oct,d} \tag{3.8.5a}$$

or also

$$\mathbf{t}_{oct} = \boldsymbol{\sigma}_{oct} + \boldsymbol{\tau}_{oct} \tag{3.8.5b}$$

note that in this case the stress $\mathbf{t}_{oct,d}$ does not have a normal component; therefore, it can be written as in (3.8.5b).

If the stress vector is represented by $\mathbf{t}_{oct}$ according to an isometric perspective in the principal stress directions, the result is the projection of the stress vector over the octahedral plane; therefore, only the tangential component will be seen over this plane, which is $\boldsymbol{\tau}_{oct}$, as $\boldsymbol{\sigma}_{oct}$ is reduced to a point on the coordinates origin. This approach is called the Haigh–Westergaard projection and appears in figure 3.8.2a. The plane defined by the vectors $\boldsymbol{\sigma}_{oct}$ and $\boldsymbol{\tau}_{oct}$ is called the *meridian plane*.

The moduli of $\boldsymbol{\sigma}_{oct}$ and $\boldsymbol{\tau}_{oct}$ and the orientation of the latter define a coordinate system $(\sigma_{oct}, \tau_{oct}, \theta)$. Angle θ is the one that forms the meridian plane with the plane defined by $\boldsymbol{\sigma}_{oct}$ and σ_1 axis. This is the coordinate system

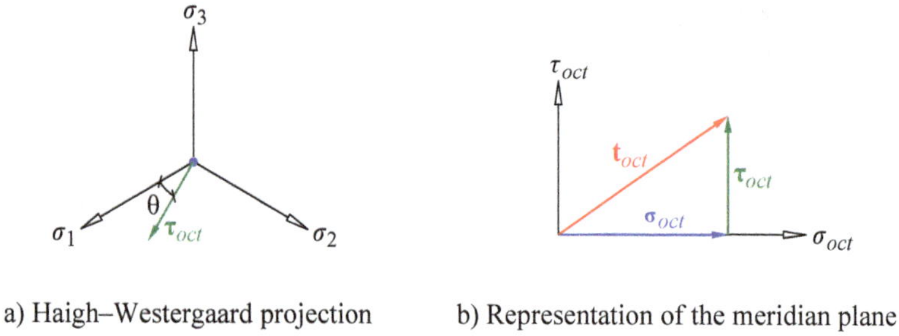

Figure 3.8.2: Haigh–Westergaard system of coordinates

of Haigh–Westergaard, and its relationship to the trihedron formed by the principal directions is

$$\begin{bmatrix} \sigma_1 \\ \sigma_2 \\ \sigma_3 \end{bmatrix} = \frac{1}{\sqrt{3}} \begin{bmatrix} \sigma_{oct} \\ \sigma_{oct} \\ \sigma_{oct} \end{bmatrix} + \sqrt{\frac{2}{3}} \tau_{oct} \begin{bmatrix} \cos\theta \\ \cos\left(\theta - \frac{2\pi}{3}\right) \\ \cos\left(\theta + \frac{2\pi}{3}\right) \end{bmatrix} \tag{3.8.6}$$

If a stress field like the one represented in figure 3.7.1 is decomposed into the two expressions of (3.8.3), the Mohr's circles of the spherical tensor are reduced to a point, defined by the coordinates $(\sigma_m, 0)$, and the ones of the deviatoric tensor are those of the stress field modified by a magnitude σ_m on the horizontal coordinate axis.

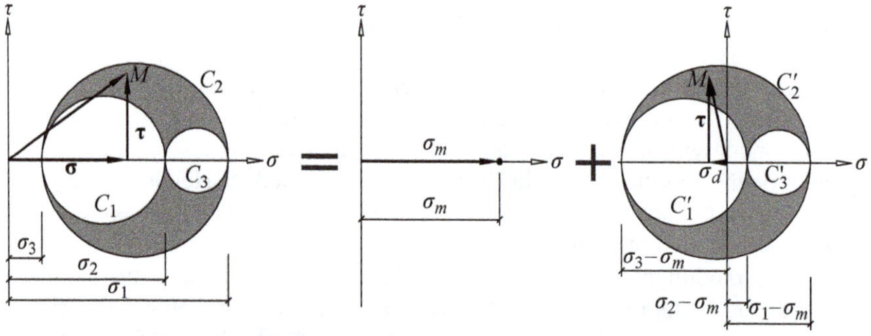

Figure 3.8.3: Mohr's circles of the spherical and deviatoric stress vectors

In the spherical tensor, the invariants present the following attributes

$$J_{1m} = J_{1oct} = 3 \cdot \sigma_m = \sigma_1 + \sigma_2 + \sigma_3 = J_1 \tag{3.8.7a}$$

$$J_{2m} = J_{2oct} = 3 \cdot \sigma_m^2 = \frac{(\sigma_1 + \sigma_2 + \sigma_3)^2}{3} = \frac{J_1^2}{3} \tag{3.8.7b}$$

$$J_{3m} = J_{3oct} = \sigma_m^3 = \frac{(\sigma_1 + \sigma_2 + \sigma_3)^3}{27} = \frac{J_1^3}{27} \tag{3.8.7c}$$

In the deviatoric tensor, the principal directions are the same as the total stress field, the principal stresses are modified by the σ_m, and the invariants have the following values

$$J_{1d} = (\sigma_1 - \sigma_m) + (\sigma_2 - \sigma_m) + (\sigma_3 - \sigma_m) = 0 \tag{3.8.8a}$$

$$J_{2d} = (\sigma_1 - \sigma_m)(\sigma_2 - \sigma_m) + (\sigma_1 - \sigma_m)(\sigma_3 - \sigma_m) + (\sigma_2 - \sigma_m)(\sigma_3 - \sigma_m) = J_2 - \frac{J_1^2}{3} \tag{3.8.8b}$$

or also

$$J_{2d} = -\frac{1}{6}\left[(\sigma_1 - \sigma_2)^2 + (\sigma_1 - \sigma_3)^2 + (\sigma_2 - \sigma_3)^2\right] \tag{3.8.8c}$$

$$J_{3d} = |\mathbf{\tau}_d| = \frac{2J_1^3}{27} - \frac{J_1 J_2}{3} + J_3 \tag{3.8.8d}$$

EXERCISES

3.1. Given a stress field, demonstrate that:

1. The octahedral stress σ_m corresponds to the normal component of a plane defined as

$$\mathbf{n} = \left[\pm 1/\sqrt{3} \quad \pm 1/\sqrt{3} \quad \pm 1/\sqrt{3}\right]^T$$

2. A stress field and its deviatoric tensor have the same principal directions.
3. If one calls t_1, t_2, t_3 the principal stresses of the deviatoric tensor, its relationship to the principal stress $\sigma_i\ (i = 1,2,3)$ of a stress field is given by

$$t_1 = \frac{2\sigma_1 - \sigma_2 - \sigma_3}{3} \qquad t_2 = \frac{2\sigma_2 - \sigma_1 - \sigma_3}{3} \qquad t_3 = \frac{2\sigma_3 - \sigma_1 - \sigma_2}{3}$$

3.2. Given the stress field:

$$\sigma_x = -4 \quad \sigma_y = 2 \quad \sigma_z = 1$$
$$\tau_{xy} = 4 \quad \tau_{xz} = 0 \quad \tau_{yz} = 0$$

1. Determine the principal stresses and their directions.
2. Find the components of the stress that act on a plane defined by a direction **n** that forms equal angles with the coordinate axes.
3. Obtain analytically the shear and normal components of the stress on that plane, and graphically using Mohr's circles.
4. Find the value of the maximum shear stress and its associated planes.
5. Obtain the surface generated by directions **n** of the planes in which the stress is equal to zero, and indicate the domain of the space in which it is positive and the domain in which it is negative.

3.3. A stress field produces the following stress tensor at a point P:

$$\boldsymbol{\tau} = \begin{bmatrix} 3 & 1 & 1 \\ 1 & 0 & 2 \\ 1 & 2 & 0 \end{bmatrix}$$

1. Obtain the moduli and directions of the normal and shear stress in the plane defined by the direction $\mathbf{n} = \left[1/\sqrt{5}\ 0\ 2/\sqrt{5}\right]^T$.
2. Obtain the invariants, principal stresses and principal directions at that point.
3. Obtain the expression for the stress quadrics.
4. Draw Mohr's circles.
5. Obtain the deviatoric and spherical tensors at a point and, as well as the values of the principal stresses of the deviatoric tensor.

3.4. The stress field of an elastic solid is defined by the following Cauchy stress tensor:

$$\boldsymbol{\tau} = \begin{bmatrix} 3 & 0 & 0 \\ 0 & -\dfrac{1}{4} & \dfrac{3\sqrt{3}}{4} \\ 0 & \dfrac{3\sqrt{3}}{4} & \dfrac{5}{4} \end{bmatrix}$$

Find:

1. The principal stresses and principal directions.
2. The spherical and deviatoric tensors. Also, find the components of the stress that act on a plane defined by a direction **n** having the same angles with the coordinate axes of the first quadrant.
3. Using the indicatrix quadric, the stress on a plane whose direction is defined by the vector **n** that, referred to the principal directions, is

$$\mathbf{n} = \begin{bmatrix} \frac{1}{2} & \frac{\sqrt{3}}{2\sqrt{2}} & \frac{\sqrt{3}}{2\sqrt{2}} \end{bmatrix}^T$$

CHAPTER 4

RELATIONSHIP BETWEEN STRESSES AND STRAINS IN DEFORMABLE BODIES. CONSTITUTIVE EQUATIONS

4.1 BEHAVIOUR OF CONTINUOUS BODIES. CONSTITUTIVE EQUATIONS

In the preceding chapters, the relationships between the displacements of the particles of the continuum and the strains that are produced were studied, and the resulting kinematic equations obtained; furthermore, a strain tensor was defined at each point. The stress state of the material under the action of external loads was also analysed and the stress tensor associated with each point obtained in a similar way as for the strains. The equations that define the equilibrium of the differential element were also obtained.

It is now necessary to define how the material behaves in the presence of stresses and strains, i.e. the relationship between the stresses and strains caused by the external loads. The equations that define this are called *constitutive equations* and depend on the type of material that is being considered. Moreover, for the same material, these constitutive equations may be different, depending on the level of stress that is reached, or they may be modified due to the stress history of the material, that is to say, the stresses that have previously occurred.

The most interesting models for the behaviour of materials are shown in figure 4.1.1:

a) *Rigid materials*. Strains do not appear at any stress level. This is the model used in the mechanics of solid bodies.
b) *Plastic materials*. The strains grow continuously with a constant stress value. When the external actions disappear, the strain produced on the material is maintained. This phenomenon occurs in metallic and polymeric materials during the forming process.
c) *Elastic and linear materials*. The stresses are proportional to the strains and the solid recovers its initial geometry as the stresses disappear. This is the typical behaviour of steel.

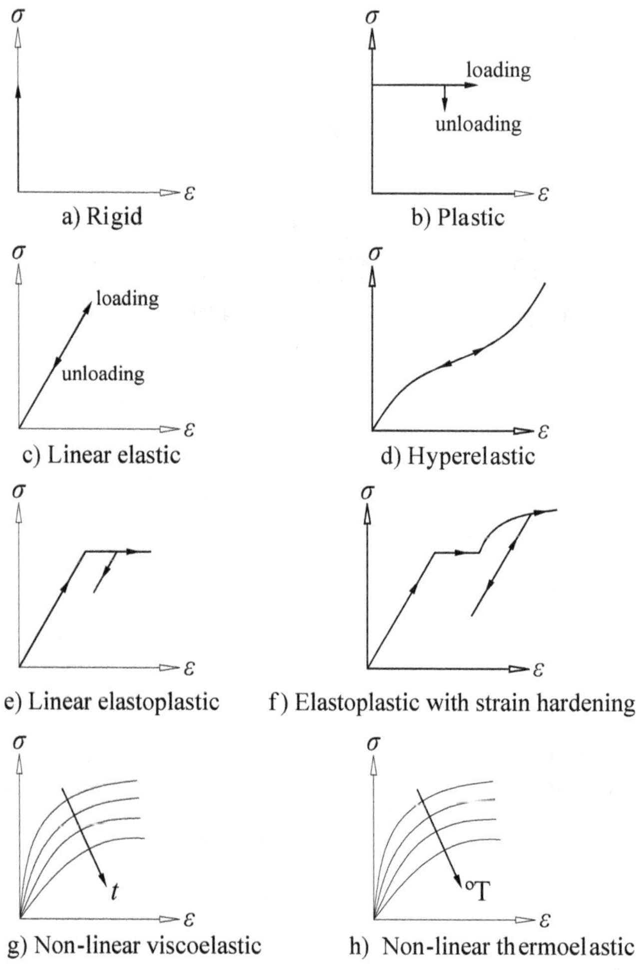

Figure 4.1.1: Models of material behaviour

d) *Hyperelastic materials*. In these materials the relationship between stresses and strains is expressed by a curved line. Concrete behaves in this way under compression. Elastomeric materials or biological tissues also have a non-linear elastic behaviour.

e) *Linear elastoplastic materials*. These materials behave as elastic and linear up to a given level of stress, and if this is exceeded, a plastic phase appears. When the material is unloaded, the strains are recovered according to a line parallel to the elastic phase.

f) *Elastoplastic materials with strain hardening*. In this case, there is an elastic phase, another plastic and then another elastic phase, as observed in figure 4.1.1f. If the required stress level is attained for the latter, the recovery of the material and a subsequent reload occurs according to a line parallel to the first elastic phase. Therefore, the reloading cycle has a longer elastic phase than that of the initial loading cycle. The structural steel is susceptible to deformation according to this model; thus, it is used in industry to obtain steels with elastic phases of different magnitudes.

g) *Viscoelastic materials.* In these materials, the strains increase over time for any stress value. Therefore, this phenomenon is time dependent and is known as material *creep*. Concrete is a practical example of non-linear viscoelastic behaviour.

h) *Thermoelastic materials.* When the degree of strain produced by the stresses depends on the temperature, the material is said to be thermoelastic. If one keeps the stress value constant, the strains increase as the temperature increases, as indicated in figure 4.1.1h. Steel becomes a special case of linear thermoelastic materials when the temperature rises to several hundreds of degrees.

Some materials may behave in a way that combines several of the above models; for example, elastomeric materials, used in bridge supports, possess hyper-visco-elastoplastic behaviour.

4.2 LINEAR ELASTIC MATERIALS. GENERALIZED HOOKE'S LAW

According to the stress tensors $\boldsymbol{\tau}$ and strain **E** previously defined, the following vectors can be written

$$\boldsymbol{\sigma} = \begin{bmatrix} \sigma_x & \sigma_y & \sigma_z & \tau_{xy} & \tau_{xz} & \tau_{yz} \end{bmatrix}^T \qquad (4.2.1)$$

$$\boldsymbol{\varepsilon} = \begin{bmatrix} \varepsilon_x & \varepsilon_y & \varepsilon_z & \frac{1}{2}\gamma_{xy} & \frac{1}{2}\gamma_{xz} & \frac{1}{2}\gamma_{yz} \end{bmatrix}^T \qquad (4.2.2)$$

where $\boldsymbol{\sigma}$ and $\boldsymbol{\varepsilon}$ are called the *stress* and *strain vectors* at a point. In linear materials, the relationship between both vectors is carried out through the *constitutive matrix* **D**, as follows

$$\boldsymbol{\sigma} = \mathbf{D}\boldsymbol{\varepsilon} \qquad (4.2.3)$$

or, in the more general form

$$\mathbf{D} = \begin{bmatrix} d_{11} & d_{12} & d_{13} & d_{14} & d_{15} & d_{16} \\ d_{21} & d_{22} & d_{23} & d_{24} & d_{25} & d_{26} \\ d_{31} & d_{32} & d_{33} & d_{34} & d_{35} & d_{36} \\ d_{41} & d_{42} & d_{43} & d_{44} & d_{45} & d_{46} \\ d_{51} & d_{52} & d_{53} & d_{54} & d_{55} & d_{56} \\ d_{61} & d_{62} & d_{63} & d_{64} & d_{65} & d_{66} \end{bmatrix} \tag{4.2.4}$$

Matrix **D** is symmetrical, and one can verify this by formulating the energy increase in the strain U per volume unit at a point with stress and strain vectors σ and ε, respectively. When the strains experience an increase $d\varepsilon$, the deformation energy change will be

$$dU = \boldsymbol{\sigma}^T d\boldsymbol{\varepsilon} \tag{4.2.5}$$

in order to facilitate the formulation, one can write the vectors of expressions (4.2.1) and (4.2.2) as

$$\boldsymbol{\sigma} = \begin{bmatrix} t_1 & t_2 & t_3 & t_4 & t_5 & t_6 \end{bmatrix}^T \tag{4.2.6}$$

$$\boldsymbol{\varepsilon} = \begin{bmatrix} \varepsilon_1 & \varepsilon_2 & \varepsilon_3 & \varepsilon_4 & \varepsilon_5 & \varepsilon_6 \end{bmatrix}^T \tag{4.2.7}$$

expression (4.2.5) then becomes

$$dU = \sum_{i=1}^{6} t_i d\varepsilon_i \tag{4.2.8}$$

since

$$dU = \sum_{i=1}^{6} \frac{\partial U}{\partial \varepsilon_i} d\varepsilon_i \tag{4.2.9}$$

this results in

$$t_i = \frac{\partial U}{\partial \varepsilon_i} \tag{4.2.10}$$

On the other hand, from equation (4.2.3), one obtains

$$t_i = \sum_{j=1}^{6} d_{ij} \varepsilon_j \tag{4.2.11}$$

then

$$\frac{\partial U}{\partial \varepsilon_i} = \sum_{j=1}^{6} d_{ij} \varepsilon_j \tag{4.2.12}$$

taking derivatives, one obtains

$$\frac{\partial^2 U}{\partial \varepsilon_i \partial \varepsilon_j} = d_{ij} \tag{4.2.13}$$

notice that

$$\frac{\partial^2 U}{\partial \varepsilon_i \partial \varepsilon_j} = \frac{\partial^2 U}{\partial \varepsilon_j \partial \varepsilon_i} \tag{4.2.14}$$

and hence it is possible to conclude that

$$d_{ij} = d_{ji} \tag{4.2.15}$$

implying that, in general

$$\mathbf{D} = \begin{bmatrix} d_{11} & & & & sym & \\ d_{21} & d_{22} & & & & \\ d_{31} & d_{32} & d_{33} & & & \\ d_{41} & d_{42} & d_{43} & d_{44} & & \\ d_{51} & d_{52} & d_{53} & d_{54} & d_{55} & \\ d_{61} & d_{62} & d_{63} & d_{64} & d_{65} & d_{66} \end{bmatrix} \tag{4.2.16}$$

Therefore, in the more general case of linear materials, that is to say, in a completely anisotropic material, there are 21 different coefficients in the constitutive matrix. That number diminishes if the elastic properties of the material present some symmetry. For instance, if symmetry occurs with respect to the *XZ* plane shown in figure 4.2.1 and the system of coordinates is changed to one also symmetric to the same plane, the new tensors $\overline{\boldsymbol{\tau}}$ and $\overline{\mathbf{E}}$ can be

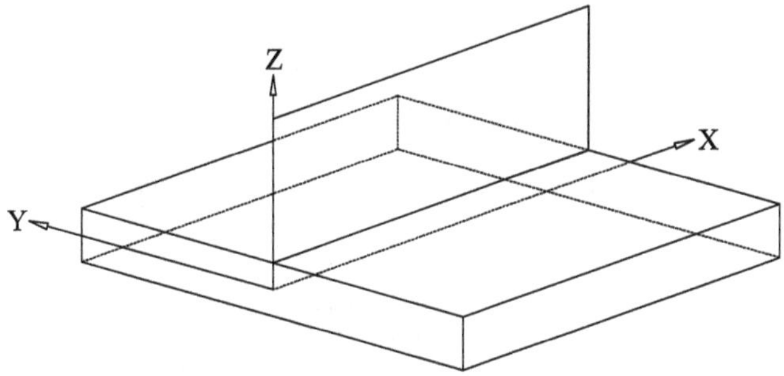

Figure 4.2.1: Material with elastic symmetry with respect to the *XZ* plane

obtained by applying the transformation matrix **T** to the following expressions

$$\mathbf{T} = \begin{bmatrix} 1 & 0 & 0 \\ 0 & -1 & 0 \\ 0 & 0 & 1 \end{bmatrix} \tag{4.2.17}$$

$$\bar{\boldsymbol{\tau}} = \mathbf{T}\boldsymbol{\tau}\mathbf{T}^T \qquad \bar{\mathbf{E}} = \mathbf{T}\mathbf{E}\mathbf{T}^T \tag{4.2.18}$$

The relationships between the elements $\bar{\boldsymbol{\tau}}$, $\bar{\mathbf{E}}$ and $\boldsymbol{\tau}$, $\mathbf{E}$ become

$$\bar{\sigma}_x = \sigma_x \qquad \bar{\sigma}_y = \sigma_y \qquad \bar{\sigma}_z = \sigma_z \tag{4.2.19a}$$

$$\bar{\tau}_{xy} = -\tau_{xy} \qquad \bar{\tau}_{xz} = \tau_{xz} \qquad \bar{\tau}_{yz} = -\tau_{yz} \tag{4.2.19b}$$

and

$$\bar{\varepsilon}_x = \varepsilon_x \qquad \bar{\varepsilon}_y = \varepsilon_y \qquad \bar{\varepsilon}_z = \varepsilon_z \tag{4.2.20a}$$

$$\bar{\gamma}_{xy} = -\gamma_{xy} \qquad \bar{\gamma}_{xz} = \gamma_{xz} \qquad \bar{\gamma}_{yz} = -\gamma_{yz} \tag{4.2.20b}$$

On the basis of (4.2.3), the expressions for σ_x and $\bar{\sigma}_x$ are

$$\sigma_x = d_{11}\varepsilon_x + d_{12}\varepsilon_y + d_{13}\varepsilon_z + d_{14}\gamma_{xy} + d_{15}\gamma_{xz} + d_{16}\gamma_{yz} \tag{4.2.21}$$

$$\bar{\sigma}_x = d_{11}\bar{\varepsilon}_x + d_{12}\bar{\varepsilon}_y + d_{13}\bar{\varepsilon}_z + d_{14}\bar{\gamma}_{xy} + d_{15}\bar{\gamma}_{xz} + d_{16}\bar{\gamma}_{yz} \tag{4.2.22}$$

applying (4.2.20) gives

$$\bar{\sigma}_x = d_{11}\varepsilon_x + d_{12}\varepsilon_y + d_{13}\varepsilon_z - d_{14}\gamma_{xy} + d_{15}\gamma_{xz} - d_{16}\gamma_{yz} \tag{4.2.23}$$

as $\sigma_x = \bar{\sigma}_x$, comparing expressions (4.2.21) and (4.2.23), one can conclude that

$$d_{14} = 0 \qquad d_{16} = 0 \tag{4.2.24}$$

carrying out the same operation on the other elements of the stress tensor gives

$$d_{24} = d_{26} = d_{34} = d_{36} = d_{45} = d_{65} = 0 \tag{4.2.25}$$

applying the condition of symmetry, the matrix **D** finally becomes

$$\mathbf{D} = \begin{bmatrix} d_{11} & & & & & \\ d_{21} & d_{22} & & & sym & \\ d_{31} & d_{32} & d_{33} & & & \\ 0 & 0 & 0 & d_{44} & & \\ d_{51} & d_{52} & d_{53} & 0 & d_{55} & \\ 0 & 0 & 0 & d_{64} & 0 & d_{66} \end{bmatrix} \tag{4.2.26}$$

so that the number of different elements of matrix **D** is reduced to 13.

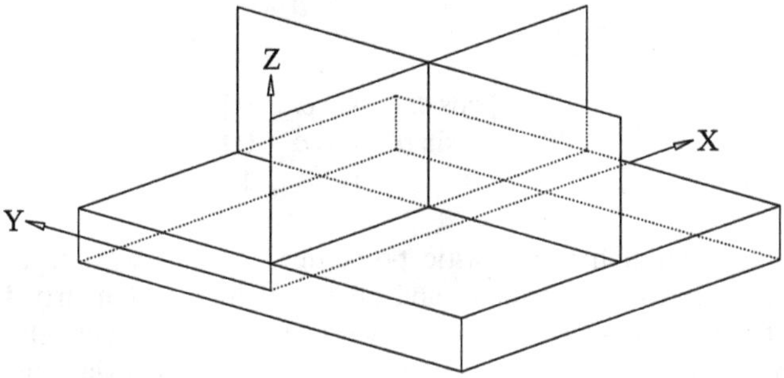

Figure 4.2.2: Material with elastic symmetry with respect to the *XZ* and *YZ* planes

If the material also has a second plane of symmetry, for example the *YZ* plane, the transformation matrix would be defined by another matrix **T** of the form

$$\mathbf{T} = \begin{bmatrix} -1 & 0 & 0 \\ 0 & 1 & 0 \\ 0 & 0 & 1 \end{bmatrix} \tag{4.2.27}$$

the resulting material is known as an *orthotropic material* and the constitutive matrix becomes

$$\mathbf{D} = \begin{bmatrix} d_{11} & & & & & \\ d_{21} & d_{22} & & & sym & \\ d_{31} & d_{32} & d_{33} & & & \\ 0 & 0 & 0 & d_{44} & & \\ 0 & 0 & 0 & 0 & d_{55} & \\ 0 & 0 & 0 & 0 & 0 & d_{66} \end{bmatrix} \tag{4.2.28}$$

The number of different elastic elements is reduced to nine.

The transformation matrices related to a rotation angle α with regard to each of the coordinate axes $\mathbf{T}_x$, $\mathbf{T}_y$, $\mathbf{T}_z$ are

$$\mathbf{T}_x = \begin{bmatrix} 1 & 0 & 0 \\ 0 & \cos\alpha & \sin\alpha \\ 0 & -\sin\alpha & \cos\alpha \end{bmatrix} \tag{4.2.29a}$$

$$\mathbf{T}_y = \begin{bmatrix} \cos\alpha & 0 & -\sin\alpha \\ 0 & 1 & 0 \\ \sin\alpha & 0 & \cos\alpha \end{bmatrix} \tag{4.2.29b}$$

$$\mathbf{T}_z = \begin{bmatrix} \cos\alpha & \sin\alpha & 0 \\ -\sin\alpha & \cos\alpha & 0 \\ 0 & 0 & 1 \end{bmatrix} \tag{4.2.29c}$$

When a material has its elastic properties constant with regard to any orientation, it is called *isotropic*, and all the elements of matrix **D** can be obtained from two constants. One can verify this by consecutively applying generic rotations with respect to each of the three axes, using the matrices defined in (4.2.29) and proceeding as in the case of planes of symmetry. After this, one can apply the condition that the elastic constants are identical for any direction, and one can conclude that the only elements of **D** not equal to zero are those that appear below, with the following relationships between them, i.e.

$$d_{12} = d_{13} = d_{21} = d_{23} = d_{31} = d_{32} = \lambda \tag{4.2.30a}$$

$$d_{44} = d_{55} = d_{66} = 2G \tag{4.2.30b}$$

$$d_{11} = d_{22} = d_{33} = d_{12} + d_{44} = \lambda + 2G \tag{4.2.30c}$$

thus, **D** results in

$$\mathbf{D} = \begin{bmatrix} \lambda+2G & & & & & \\ \lambda & \lambda+2G & & & sym & \\ \lambda & \lambda & \lambda+2G & & & \\ 0 & 0 & 0 & 2G & & \\ 0 & 0 & 0 & 0 & 2G & \\ 0 & 0 & 0 & 0 & 0 & 2G \end{bmatrix} \tag{4.2.31}$$

In structural engineering, it is common to denote vector $\boldsymbol{\varepsilon}$ defined in (4.2.2) in the following way

$$\boldsymbol{\varepsilon} = \begin{bmatrix} \varepsilon_x & \varepsilon_y & \varepsilon_z & \gamma_{xy} & \gamma_{xz} & \gamma_{yz} \end{bmatrix}^T \tag{4.2.32}$$

as a result, the constitutive matrix that relates the vectors $\boldsymbol{\sigma}$, $\boldsymbol{\varepsilon}$ is

$$\mathbf{D} = \begin{bmatrix} \lambda+2G & & & & & \\ \lambda & \lambda+2G & & & sym & \\ \lambda & \lambda & \lambda+2G & & & \\ 0 & 0 & 0 & G & & \\ 0 & 0 & 0 & 0 & G & \\ 0 & 0 & 0 & 0 & 0 & G \end{bmatrix} \tag{4.2.33}$$

from this, one can obtain the following relationships, known as *Lame's equations*.

$$\sigma_x = 2G\varepsilon_x + \lambda e \quad (4.2.34a)$$

$$\sigma_y = 2G\varepsilon_y + \lambda e \quad (4.2.34b)$$

$$\sigma_z = 2G\varepsilon_z + \lambda e \quad (4.2.34c)$$

$$\tau_{xy} = G\gamma_{xy} \qquad \tau_{xz} = G\gamma_{xz} \qquad \tau_{yz} = G\gamma_{yz} \quad (4.2.34d)$$

where e, the *volumetric strain*, is defined as

$$e = \varepsilon_x + \varepsilon_y + \varepsilon_z \quad (4.2.35)$$

In most cases, experimental measurements give two mechanical parameters corresponding to the *longitudinal strain modulus E*, known as *Young's modulus* and the *transverse strain coefficient* ν or *Poisson's coefficient*, which are associated with λ and G in the following way

$$\lambda = \frac{\nu E}{(1+\nu)(1-2\nu)} \qquad G = \frac{E}{2(1+\nu)} \quad (4.2.36)$$

as a result, one can write matrix **D** as

$$\mathbf{D} = \frac{E}{(1+\nu)(1-2\nu)} \begin{bmatrix} 1-\nu & & & & & \\ \nu & 1-\nu & & & sym & \\ \nu & \nu & 1-\nu & & & \\ 0 & 0 & 0 & 0.5-\nu & & \\ 0 & 0 & 0 & 0 & 0.5-\nu & \\ 0 & 0 & 0 & 0 & 0 & 0.5-\nu \end{bmatrix} \quad (4.2.37a)$$

or also as

$$\mathbf{D} = \frac{G}{0.5-\nu} \begin{bmatrix} 1-\nu & & & & & \\ \nu & 1-\nu & & & sym & \\ \nu & \nu & 1-\nu & & & \\ 0 & 0 & 0 & 0.5-\nu & & \\ 0 & 0 & 0 & 0 & 0.5-\nu & \\ 0 & 0 & 0 & 0 & 0 & 0.5-\nu \end{bmatrix} \quad (4.2.37b)$$

Alternatively, one can express these equations inversely, giving rise to what is known as the *generalized Hooke's law*, i.e.

$$\varepsilon_x = \frac{1}{E}\left[\sigma_x - \nu\left(\sigma_y + \sigma_z\right)\right] \tag{4.2.38a}$$

$$\varepsilon_y = \frac{1}{E}\left[\sigma_y - \nu\left(\sigma_x + \sigma_z\right)\right] \tag{4.2.38b}$$

$$\varepsilon_z = \frac{1}{E}\left[\sigma_z - \nu\left(\sigma_x + \sigma_y\right)\right] \tag{4.2.38c}$$

similarly

$$\gamma_{xy} = \frac{\tau_{xy}}{G} \qquad \gamma_{xz} = \frac{\tau_{xz}}{G} \qquad \gamma_{yz} = \frac{\tau_{yz}}{G} \tag{4.2.38d}$$

In the directions of the principal stresses equations (4.2.36), these become

$$\varepsilon_1 = \frac{1}{E}\left[\sigma_1 - \nu\left(\sigma_2 + \sigma_3\right)\right] \tag{4.2.39a}$$

$$\varepsilon_2 = \frac{1}{E}\left[\sigma_2 - \nu\left(\sigma_1 + \sigma_3\right)\right] \tag{4.2.39b}$$

$$\varepsilon_3 = \frac{1}{E}\left[\sigma_3 - \nu\left(\sigma_1 + \sigma_2\right)\right] \tag{4.2.39c}$$

$$\gamma_{xy} = \gamma_{xz} = \gamma_{yz} = 0 \tag{4.2.39d}$$

Equation (4.2.38) shows that the tensile stresses produce extensions in the direction in which they act and shortenings in perpendicular directions. For example, if in a continuum the only principal stress not equal to zero is σ_1, the resulting strains will be

$$\varepsilon_1 = \frac{\sigma_1}{E} \qquad \varepsilon_2 = -\nu\frac{\sigma_1}{E} \qquad \varepsilon_3 = -\nu\frac{\sigma_1}{E} \tag{4.2.40}$$

therefore, the volume increase in a differential element will be

$$\frac{dV}{V} = \frac{dxdydz\left(1 + \varepsilon_1\right)\left(1 - \nu\varepsilon_1\right)^2 - dxdydz}{dxdydz} \tag{4.2.41}$$

which can be approximated by the expression

$$\frac{dV}{V} \approx \varepsilon_1\left(1 - 2\nu\right) \tag{4.2.42}$$

Most materials experience an increase in volume under traction forces, which means that dV must be positive; thus, the following condition must be met

$$\nu < 0.5 \tag{4.2.43}$$

The volumetric strain e defined above can be expressed using (4.2.38) as

$$e = \frac{(\sigma_1 + \sigma_2 + \sigma_3)(1 - 2\nu)}{E} = \frac{J_1 (1 - 2\nu)}{E} \tag{4.2.44}$$

where J_1 is the linear invariant of stress previously defined.

If the continuum is in a hydrostatic stress field that is expressed as

$$\sigma_x = \sigma_y = \sigma_z = -p \tag{4.2.45}$$

the volumetric strain becomes

$$e = \frac{-3p(1 - 2\nu)}{E} = \frac{-p}{K} \tag{4.2.46}$$

where

$$K = \frac{E}{3(1 - 2\nu)} \tag{4.2.47}$$

and the parameter K is commonly known as the *bulk elasticity modulus.*

4.3 STRAIN ENERGY

Let us consider a homogeneous and isotropic solid in an elastic and linear system subjected to external loads in which the normal stress and the longitudinal strain in the direction **n** have achieved the values σ and ε, respectively. During the strain process, intermediate values will exist as $\alpha\sigma$ and $\alpha\varepsilon$, where $0 \leq \alpha \leq 1$.

For each infinitesimal strain increase, the stresses will increase similarly, and the increase in strain energy per unit volume in direction **n** will be

$$dU = [\alpha\sigma + d(\alpha\sigma)]\, d(\alpha\varepsilon) = (\alpha\sigma + \sigma d\alpha)\,\varepsilon d\alpha = \sigma\varepsilon\alpha d\alpha + \sigma\varepsilon d\alpha^2 \approx \sigma\varepsilon\alpha d\alpha \tag{4.3.1}$$

the total strain energy throughout the process along the direction **n** will be expressed as

$$U = \int_0^1 \sigma\varepsilon\alpha d\alpha = \frac{1}{2}\sigma\varepsilon \tag{4.3.2}$$

which is equal to the area of the triangle under the line in figure 4.3.1 that relates the stresses and strains in an elastic solid.

In general, for a set of stresses and strains, the expression for the variation in the strain energy is

$$dU = \alpha\boldsymbol{\tau} : d(\alpha\mathbf{E}) = (\boldsymbol{\tau} : \mathbf{E})\,\alpha d\alpha \tag{4.3.3}$$

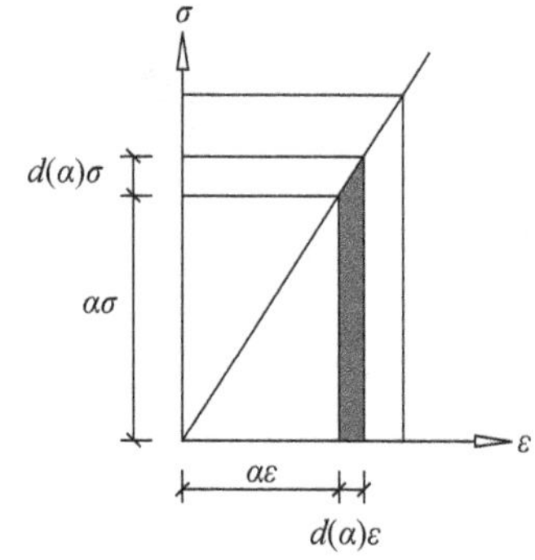

Figure 4.3.1: Strain process in linear elasticity

and the total strain energy throughout the process is

$$U = \int_0^1 (\mathbf{T}:\mathbf{E})\,\alpha d\alpha = (\mathbf{T}:\mathbf{E})\int_0^1 \alpha d\alpha = \frac{1}{2}\mathbf{T}:\mathbf{E} \tag{4.3.4}$$

This expression can be written in Cartesian axes as

$$U = \frac{1}{2}\left(\sigma_x\varepsilon_x + \sigma_y\varepsilon_y + \sigma_z\varepsilon_z + \tau_{xy}\gamma_{xy} + \tau_{xz}\gamma_{xz} + \tau_{yz}\gamma_{yz}\right) \tag{4.3.5}$$

and in the axes corresponding to the principal directions it becomes

$$U = \frac{1}{2}\sigma_1\varepsilon_1 + \frac{1}{2}\sigma_2\varepsilon_2 + \frac{1}{2}\sigma_3\varepsilon_3 \tag{4.3.6}$$

One can write this formulae based only on the stresses using expression (4.2.38), and in this case it becomes

$$U = \frac{1}{2E}\left[\sigma_x^2 + \sigma_y^2 + \sigma_z^2 - 2\nu\left(\sigma_x\sigma_y + \sigma_x\sigma_z + \sigma_y\sigma_z\right)\right] + \frac{1}{2G}\left(\tau_{xy}^2 + \tau_{xz}^2 + \tau_{yz}^2\right) \tag{4.3.7}$$

If the axes are those of the principal directions and expression (4.2.39) is used, one obtains

$$U = \frac{1}{2E}\left[\sigma_1^2 + \sigma_2^2 + \sigma_3^2 - 2\nu\left(\sigma_1\sigma_2 + \sigma_1\sigma_3 + \sigma_2\sigma_3\right)\right] \tag{4.3.8}$$

The strain energy can be similarly expressed, based on the strains, and turns out to be

$$U = \frac{G}{2}\left[2\left(\varepsilon_x^2 + \varepsilon_y^2 + \varepsilon_z^2\right) + \frac{2\nu}{1-2\nu}\left(\varepsilon_x + \varepsilon_y + \varepsilon_z\right)^2 + \gamma_{xy}^2 + \gamma_{xz}^2 + \gamma_{yz}^2\right] \tag{4.3.9}$$

In terms of principal strain directions, one finds

$$U = \frac{G}{2}\left[2\left(\varepsilon_1^2 + \varepsilon_2^2 + \varepsilon_3^2\right) + \frac{2\nu}{1-2\nu}(\varepsilon_1 + \varepsilon_2 + \varepsilon_3)^2\right] \tag{4.3.10}$$

EXERCISES

4.1. A prismatic element has undergone an increase in temperature of ΔT. Calculate the resulting strains and stresses in the following cases:

1. No displacements are allowed along one coordinate axis.
2. No displacements are allowed along two coordinate axes.
3. No displacements are allowed along three coordinate axes.

Particularize for these values:

$$E = 2.1 \cdot 10^5 \text{ MPa} \quad \nu = 0.15 \quad \Delta T = 35°\text{C} \quad \alpha_T = 0.000012 \text{ m/m°C}$$

4.2. A bar with a rectangular section $a = 100$ mm, $b = 50$ mm and length $l = 2$ m is subjected to a force $N = 50$ kN and experiences an extension $\Delta l = 1$ mm and a lateral contraction $\Delta b = -0.007$ mm. Evaluate:

1. The longitudinal elasticity modulus of the bar.
2. Poisson's coefficient.
3. The variation in the lateral dimension a of the bar.

4.3. Demonstrate that one can write the equations of internal equilibrium for an elastic and isotropic material in the following way:

$$(\lambda + G)\,\nabla e + G\nabla^2\mathbf{u} + \mathbf{b} - \rho\mathbf{a} = \mathbf{0}$$

where λ, G are the material's elastic parameters, e is the volumetric strain parameter, and **u**, **b**, **a** are the vectors of displacement, volume and acceleration forces.

4.4. Assuming that in a linear and elastic material the points have the same elastic properties for any rotation with regard to X axis, find how to simplify the elements of the constitutive matrix **D**. Remember that the matrix that defines an arbitrary rotation with respect to X axis is

$$\mathbf{T} = \begin{bmatrix} 1 & 0 & 0 \\ 0 & \cos\alpha & \sin\alpha \\ 0 & -\sin\alpha & \cos\alpha \end{bmatrix}$$

4.5. The initial sizing of a neoprene support device for a bridge, such as the one shown in the figure below, has the dimensions $1 \times 1 \times 0.1$ m, and is under a vertical load of 20 MN. Given that this value is excessive for this material, it has been decided to confine it in a very thick steel box that can be considered to be infinitely rigid; thus, the neoprene can only be deformed in a vertical direction. The mechanical characteristics of the neoprene are

$$E = 2000 \text{ MPa} \qquad \nu = 0.45$$

Find:

1. The values of the resulting stresses σ_x, σ_y.
2. The value of the neoprene's vertical displacement and compare it with what would have been obtained if the neoprene had not been confined.
3. The values of the normal and shear stresses at the points of a plane containing the z axis and the straight line AB.

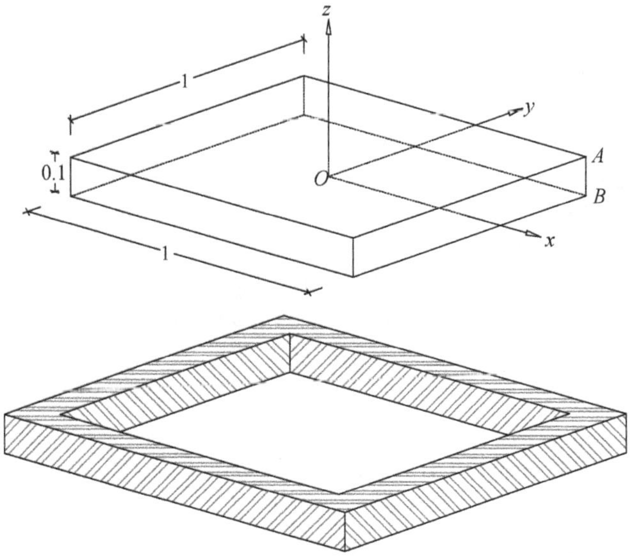

CHAPTER 5

PLANE LINEAR ELASTICITY. PLANE STRAIN AND PLANE STRESS

5.1 PLANE STRAIN FIELD

A *plane strain field* is defined by a displacements vector **u** of the form

$$\mathbf{u} = \begin{bmatrix} u(x,y) \\ v(x,y) \\ 0 \end{bmatrix} \tag{5.1.1}$$

that is to say, the displacements u, v depend on the coordinates x, y and the displacement w is equal to zero in the direction of the z axis.

Consequently, this gives

$$\gamma_{xz} = \frac{\partial u}{\partial z} + \frac{\partial w}{\partial x} = 0 \quad \gamma_{yz} = \frac{\partial v}{\partial z} + \frac{\partial w}{\partial y} = 0 \quad \varepsilon_z = \frac{\partial w}{\partial z} = 0 \tag{5.1.2}$$

and therefore

$$\tau_{xz} = \tau_{yz} = 0 \quad \sigma_z = \nu\left(\sigma_x + \sigma_y\right) \tag{5.1.3}$$

observing (5.1.2), it is clear that one of the principal stresses is σ_z, and the other two will be in the xy plane.

The stress tensor $\boldsymbol{\tau}$ is reduced to

$$\boldsymbol{\tau} = \begin{bmatrix} \sigma_x & \tau_{xy} & 0 \\ \tau_{xy} & \sigma_y & 0 \\ 0 & 0 & \sigma_z \end{bmatrix} \tag{5.1.4}$$

and Mohr's circles of stress become as shown in figure 5.1.1.

However, as the stress σ_z depends on σ_x and σ_y, the stress tensor is often written in a reduced way as

$$\boldsymbol{\tau} = \begin{bmatrix} \sigma_x & \tau_{xy} \\ \tau_{xy} & \sigma_y \end{bmatrix} \tag{5.1.5}$$

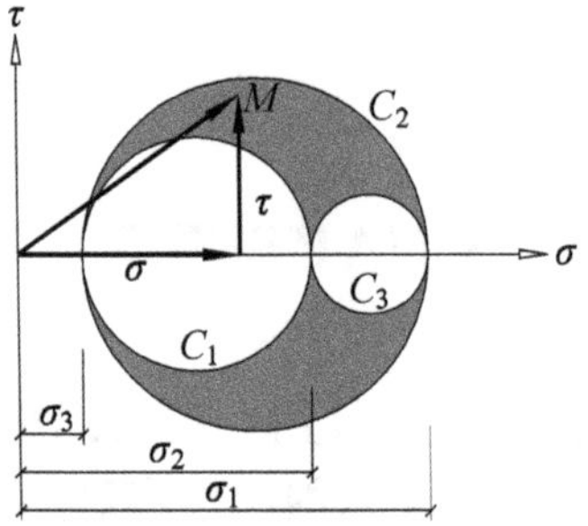

Figure 5.1.1: Mohr's circles of stress in plane strain

suppose that there exists an elastic problem defined by the following displacements

$$\mathbf{u} = \begin{bmatrix} u(x,y) \\ v(x,y) \\ az + b \end{bmatrix} \tag{5.1.6}$$

where a and b are constant values. In this case, equations (5.1.2) become

$$\gamma_{xz} = 0 \quad \gamma_{yz} = 0 \quad \varepsilon_z = a \tag{5.1.7}$$

as

$$\varepsilon_z = \frac{1}{E}\left[\sigma_z - \nu\left(\sigma_x + \sigma_y\right)\right] \tag{5.1.8}$$

$$\sigma_z = E\varepsilon_z + \nu\left(\sigma_x + \sigma_y\right) = Ea + \nu\left(\sigma_x + \sigma_y\right) \tag{5.1.9}$$

Then, such a strain field can be studied as the superposition of a plane strain field and a uniform compression or extension in the z direction.

As mentioned above, the plane strain field needs the displacement on the z axis to be equal to zero, which is the same as saying that both ends of the solid in that direction experience no displacements. Although this rarely occurs, there are situations in which the displacement w is much smaller than the other two components of the displacement vector **u**. This happens in tunnels, retaining walls, gravity dams and pressure pipes. The stress field of those structures can be analysed as plane strain fields (figure 5.1.2).

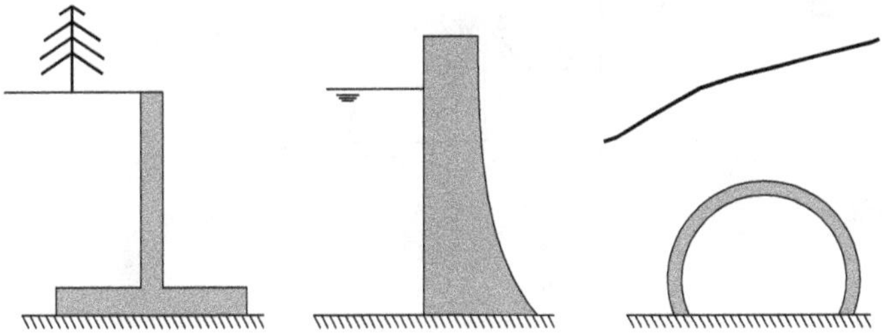

Figure 5.1.2: Examples of cases that can be represented as plane strain

The equations of equilibrium become

$$\frac{\partial \sigma_x}{\partial x} + \frac{\partial \tau_{xy}}{\partial y} + b_x - \rho a_x = 0 \tag{5.1.10a}$$

$$\frac{\partial \tau_{xy}}{\partial x} + \frac{\partial \sigma_y}{\partial y} + b_y - \rho a_y = 0 \tag{5.1.10b}$$

To obtain the expressions for Hooke's law in this case, one has to recall that

$$\varepsilon_z = 0 \quad \sigma_z = \nu\left(\sigma_x + \sigma_y\right) \tag{5.1.11}$$

then

$$\varepsilon_x = \frac{1}{E}\left[\sigma_x - \nu\left(\sigma_y + \nu\sigma_x + \nu\sigma_y\right)\right] = \frac{1+\nu}{E}\left[(1-\nu)\,\sigma_x - \nu\sigma_y\right] \tag{5.1.12a}$$

$$\varepsilon_y = \frac{1}{E}\left[\sigma_y - \nu\left(\sigma_x + \nu\sigma_x + \nu\sigma_y\right)\right] = \frac{1+\nu}{E}\left[(1-\nu)\,\sigma_y - \nu\sigma_x\right] \tag{5.1.12b}$$

$$\gamma_{xy} = \frac{\tau_{xy}}{G} \quad \gamma_{xz} = \gamma_{yz} = 0 \tag{5.1.12c}$$

The compatibility conditions are reduced to

$$\frac{\partial^2 \gamma_{xy}}{\partial x \partial y} = \frac{\partial^2 \varepsilon_y}{\partial x^2} + \frac{\partial^2 \varepsilon_x}{\partial y^2} \tag{5.1.13}$$

5.2 PLANE STRESS FIELD

A stress field in which one of the principal stresses is zero is defined as a *plane stress field*. Assuming that this happens in the z axis, this means that

$$\sigma_z = \tau_{xz} = \tau_{yz} = 0 \tag{5.2.1}$$

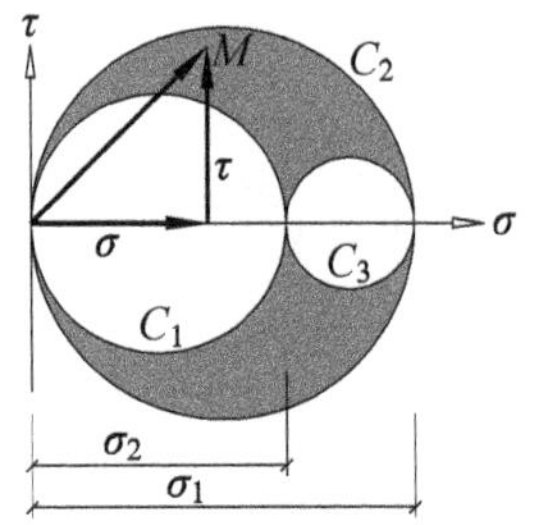

Figure 5.2.1: Mohr's circles of stress for plane stress

consequently, the stress tensor $\mathbf{T}$ is reduced to

$$\mathbf{T} = \begin{bmatrix} \sigma_x & \tau_{xy} \\ \tau_{xy} & \sigma_y \end{bmatrix} \tag{5.2.2}$$

Mohr's circles of stress for this case are shown in figure 5.2.1.

The internal equilibrium equations are the same as those for plane strain, as given by expression (5.1.10).

The expressions for Hooke's law become

$$\varepsilon_x = \frac{1}{E}\left(\sigma_x - \nu\sigma_y\right) \quad \varepsilon_y = \frac{1}{E}\left(\sigma_y - \nu\sigma_x\right) \quad \varepsilon_z = -\frac{\nu}{E}\left(\sigma_x + \sigma_y\right) \tag{5.2.3a}$$

$$\gamma_{xy} = \frac{\tau_{xy}}{G} \qquad \gamma_{xz} = \gamma_{yz} = 0 \tag{5.2.3b}$$

The compatibility equations are reduced to

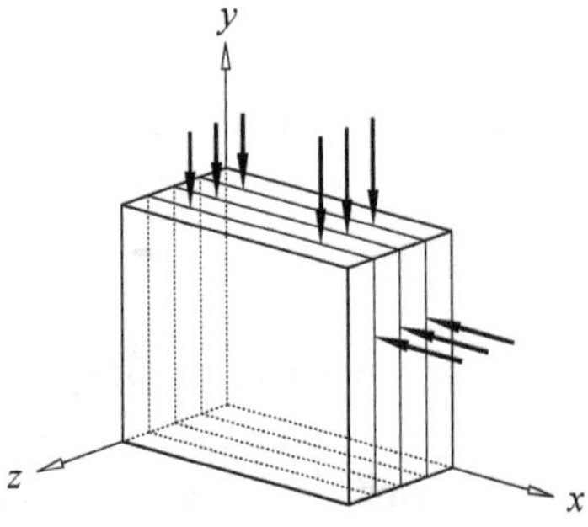

Figure 5.2.2: Stress field similar to plane stress

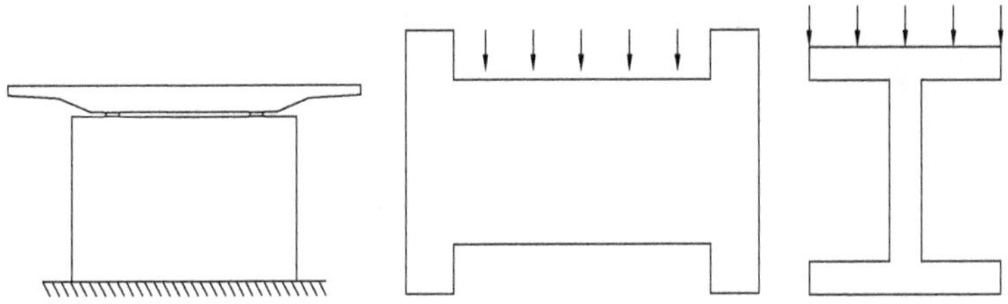

a) Bridge deck support b) Very thick beam c) Steel profile

Figure 5.2.3: Structural examples similar to plane stress fields

$$\frac{\partial^2 \varepsilon_x}{\partial y^2} + \frac{\partial^2 \varepsilon_y}{\partial x^2} = \frac{\partial^2 \gamma_{xy}}{\partial x \partial y} \tag{5.2.4a}$$

$$\frac{\partial^2 \varepsilon_z}{\partial x^2} = \frac{\partial^2 \varepsilon_z}{\partial y^2} = \frac{\partial^2 \varepsilon_z}{\partial x \partial y} = 0 \tag{5.2.4b}$$

Real-life examples of structural problems that are similar to plane stress fields are those in which one of the dimensions of the deformable solid body is much smaller than the others and the loads are contained in planes parallel to the width of the smaller dimension. Figures 5.2.3a and 5.2.3b represent a bridge deck support and a very thick concrete beam, respectively. In these examples as well as in the web of the wide flange steel profiles, the stress field can be represented as plane stress.

5.3 PRINCIPAL DIRECTIONS IN PLANE ELASTICITY

In plane strain fields just as in plane stress fields, one of the principal directions is the z axis. As indicated in (5.2.1), the plane stress case gives

$$\sigma_z = \tau_{xy} = \tau_{xz} = 0 \tag{5.3.1}$$

and in plane strain becomes

$$\sigma_z = \nu\left(\sigma_x + \sigma_y\right) \qquad \tau_{xz} = \tau_{yz} = 0 \tag{5.3.2}$$

consequently, it is sufficient in the two cases to study the stress tensor $\boldsymbol{\tau}$ reduced to

$$\boldsymbol{\tau} = \begin{bmatrix} \sigma_x & \tau_{xy} \\ \tau_{xy} & \sigma_y \end{bmatrix} \tag{5.3.3}$$

The other two principal directions are obtained from the equation

$$\begin{vmatrix} \sigma_x - \sigma & \tau_{xy} \\ \tau_{xy} & \sigma_y - \sigma \end{vmatrix} = 0 \tag{5.3.4}$$

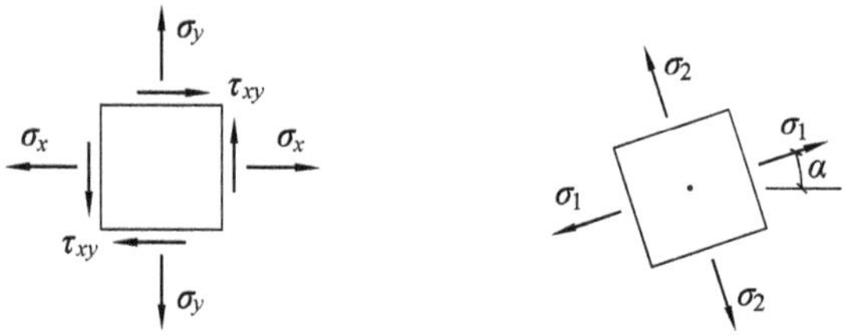

a) Stresses along the coordinate axes b) Principal directions and stresses

Figure 5.3.1: Two-dimensional stress field

and, as was seen in chapter 3, the principal direction values are

$$\sigma_1 = \frac{\sigma_x + \sigma_y}{2} + \sqrt{\left(\frac{\sigma_x - \sigma_y}{2}\right)^2 + \tau_{xy}^2} \quad \sigma_2 = \frac{\sigma_x + \sigma_y}{2} - \sqrt{\left(\frac{\sigma_x - \sigma_y}{2}\right)^2 + \tau_{xy}^2} \tag{5.3.5}$$

and the angle that the principal directions form with the x axis is given by

$$\tan(2\alpha) = \frac{2\tau_{xy}}{\sigma_x - \sigma_y} \tag{5.3.6}$$

which provides two values of α, separated by 90°, corresponding to the orientations of the principal directions.

5.4 MOHR'S CIRCLE IN PLANE ELASTICITY

In both plane stress and plane strain problems, the study of the stress field is defined by the components σ_x, σ_y, τ_{xy}.

If one defines a direction $\mathbf{n}$ using the vector $\mathbf{n} = [\cos\alpha \quad \sin\alpha]^T$, one can obtain the normal and shear stresses that correspond to that direction, i.e.

$$\mathbf{t}_n = \boldsymbol{\tau}\mathbf{n} = \begin{bmatrix} \sigma_x & \tau_{xy} \\ \tau_{xy} & \sigma_y \end{bmatrix} \begin{bmatrix} \cos\alpha \\ \sin\alpha \end{bmatrix} = \begin{bmatrix} \sigma_x \cos\alpha + \tau_{xy}\sin\alpha \\ \tau_{xy}\cos\alpha + \sigma_y \sin\alpha \end{bmatrix} \tag{5.4.1}$$

$$\sigma = \mathbf{n}^T \mathbf{t}_n = \sigma_x \cos^2\alpha + \sigma_y \sin^2\alpha + 2\tau_{xy}\sin\alpha\cos\alpha \tag{5.4.2}$$

$$\tau^2 = |\mathbf{t}_n|^2 - \sigma^2 = \left(\sigma_x\cos\alpha + \tau_{xy}\sin\alpha\right)^2 + \left(\tau_{xy}\cos\alpha + \sigma_y\sin\alpha\right)^2 - \left(\sigma_x\cos^2\alpha + \sigma_y\sin^2\alpha + 2\tau_{xy}\sin\alpha\cos\alpha\right)^2 \tag{5.4.3}$$

Developing the previous expression using the relationships between the trigonometric functions, one can find

$$\tau = \frac{(\sigma_y - \sigma_x)\sin(2\alpha)}{2} + \tau_{xy}\cos(2\alpha) \tag{5.4.4}$$

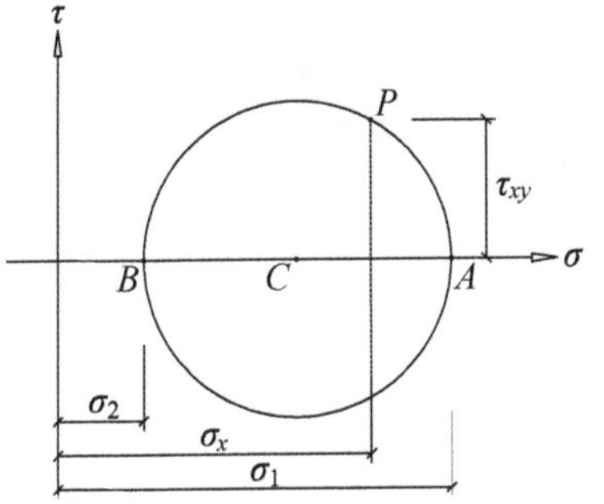

Figure 5.4.1: Mohr's circle in plane elasticity

and equation (5.4.2) can be written as

$$\sigma = \frac{\sigma_x + \sigma_y}{2} + \frac{(\sigma_x - \sigma_y)\cos(2\alpha)}{2} + \tau_{xy}\sin(2\alpha) \tag{5.4.5}$$

squaring the two expressions and regrouping the terms gives the following equation

$$\left(\sigma - \frac{\sigma_x + \sigma_y}{2}\right)^2 + \tau^2 = \left(\frac{\sigma_x - \sigma_y}{2}\right)^2 + \tau_{xy}^2 \tag{5.4.6}$$

This equation corresponds to a circle with its centre at the coordinates $[(\sigma_x + \sigma_y)/2, 0]$ and with a radius of $\sqrt{[(\sigma_x - \sigma_y)/2]^2 + \tau_{xy}^2}$ and is the locus of the stresses of the case under study, and therefore represents Mohr's circle in plane elasticity problems.

Points A and B represent the values of the principal stresses, which are given by

$$\sigma_1 = \frac{\sigma_x + \sigma_y}{2} + \sqrt{\left(\frac{\sigma_x - \sigma_y}{2}\right)^2 + \tau_{xy}^2} \tag{5.4.7a}$$

$$\sigma_2 = \frac{\sigma_x + \sigma_y}{2} - \sqrt{\left(\frac{\sigma_x - \sigma_y}{2}\right)^2 + \tau_{xy}^2} \tag{5.4.7b}$$

one can verify that these values correspond to those obtained in chapter 3 using expressions (3.4.8).

Point P, whose coordinates are (σ_x, τ_{xy}), is called the *pole* of Mohr's circle, and from it one can find out the plane that corresponds to each stress vector.

Given a stress field such as the one defined in figure 5.4.2, one can draw a horizontal line from point P, intersecting the circle at point Q of coordinates

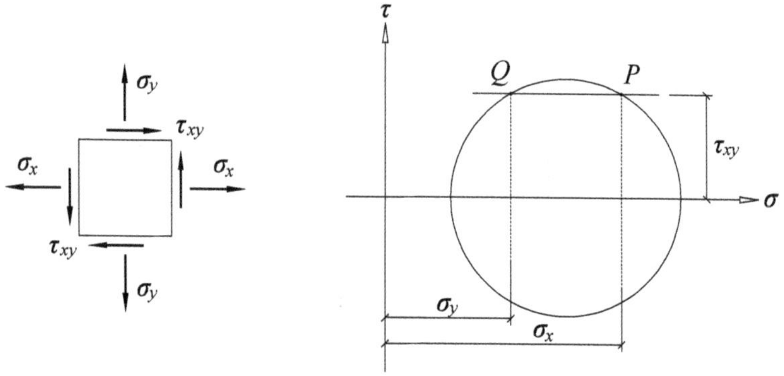

Figure 5.4.2: Stress vectors in two-dimensional problems

(σ_y, τ_{xy}), which correspond to the stress components in the horizontal plane. Consequently, by drawing any straight line from point P, the point that cuts Mohr's circle is the one that represents the stress components in the parallel plane to the straight line. This property is easy to prove for the vertical plane.

The principal stresses will be in the planes defined by the *PA* and *PB* segments in figure 5.4.3. The *PA* plane corresponds to σ_1 and the plane defined by *PB* to σ_2. The angle between σ_1 and the horizontal axis, as shown in figure 5.3.1b, can be obtained from the inclination of the *CP* segment whose value is given by

$$\tan(2\alpha) = \frac{\tau_{xy}}{\sigma_x - \dfrac{\sigma_x + \sigma_y}{2}} = \frac{2\tau_{xy}}{\sigma_x - \sigma_y} \tag{5.4.8}$$

which is the same expression as previously obtained.

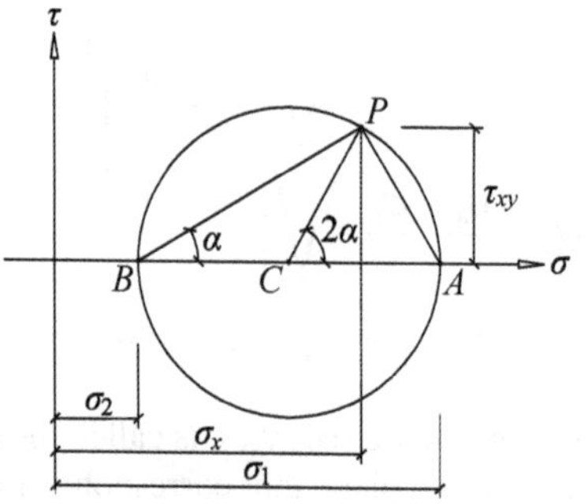

Figure 5.4.3: Orientation of the principal directions

5.5 STRESS FUNCTION IN PLANE ELASTICITY. AIRY FUNCTION

5.5.1 Plane strain field

The internal equilibrium equations have been expressed as (5.1.10). Differentiating (5.1.10a) with respect to the variable x and (5.1.10b) with respect to y gives

$$\frac{\partial^2 \sigma_x}{\partial x^2} + \frac{\partial^2 \tau_{yx}}{\partial x \partial y} + \frac{\partial}{\partial x}(b_x - \rho a_x) = 0 \tag{5.5.1.1a}$$

$$\frac{\partial^2 \tau_{xy}}{\partial x \partial y} + \frac{\partial^2 \sigma_y}{\partial y^2} + \frac{\partial}{\partial y}(b_y - \rho a_y) = 0 \tag{5.5.1.1b}$$

adding both expressions and grouping terms, one finds

$$2\frac{\partial^2 \tau_{xy}}{\partial x \partial y} = -\frac{\partial^2 \sigma_x}{\partial x^2} - \frac{\partial^2 \sigma_y}{\partial y^2} - \frac{\partial}{\partial x}(b_x - \rho a_x) - \frac{\partial}{\partial y}(b_y - \rho a_y) \tag{5.5.1.2}$$

The compatibility equation was written as

$$\frac{\partial^2 \gamma_{xy}}{\partial x \partial y} = \frac{\partial^2 \varepsilon_x}{\partial y^2} + \frac{\partial^2 \varepsilon_y}{\partial x^2} \tag{5.5.1.3}$$

One can carry out those derivatives in the expressions for Hooke's law (5.1.12) and substitute the result in (5.5.1.3) to obtain

$$2\frac{\partial^2 \tau_{xy}}{\partial x \partial y} = \frac{\partial^2 \sigma_x}{\partial y^2} + \frac{\partial^2 \sigma_y}{\partial x^2} - \nu\left(\frac{\partial^2 \sigma_x}{\partial x^2} + \frac{\partial^2 \sigma_x}{\partial y^2} + \frac{\partial^2 \sigma_y}{\partial x^2} + \frac{\partial^2 \sigma_y}{\partial y^2}\right) \tag{5.5.1.4}$$

comparing (5.5.1.2) and (5.5.1.4), it turns out that

$$\left(\frac{\partial^2}{\partial x^2} + \frac{\partial^2}{\partial y^2}\right)(\sigma_x + \sigma_y) = \frac{1}{\nu - 1}\left[\frac{\partial}{\partial x}(b_x - \rho a_x) + \frac{\partial}{\partial y}(b_y - \rho a_y)\right] \tag{5.5.1.5a}$$

using the divergence operator, the equation can be written as

$$\nabla^2(\sigma_x + \sigma_y) = \frac{1}{\nu - 1}\nabla(\mathbf{b} - \rho\mathbf{a}) \tag{5.5.1.5b}$$

If the mass and acceleration forces are constant, the right-hand side of the equation becomes zero, and in that case, one can obtain the expressions for the stress tensor components from a function called the stress function Φ, attributed to Airy, i.e.

$$\sigma_x = \frac{\partial^2 \Phi}{\partial y^2} \qquad \sigma_y = \frac{\partial^2 \Phi}{\partial x^2} \tag{5.5.1.6a}$$

$$\tau_{xy} = -\frac{\partial^2 \Phi}{\partial x \partial y} - (b_x - \rho a_x)\, y - \left(b_y - \rho a_y\right) x \tag{5.5.1.6b}$$

one can verify that the compatibility condition defined by equation (5.5.1.4) is met, and equation (5.5.1.5) results in

$$\nabla^4 \Phi = \frac{\partial^4 \Phi}{\partial x^4} + 2\frac{\partial^4 \Phi}{\partial x^2 \partial y^2} + \frac{\partial^4 \Phi}{\partial y^4} = 0 \tag{5.5.1.7}$$

5.5.2 Plane stress field

In the plane stress fields, the compatibility equations were (5.2.4). One of them has already been used in the previous section, i.e.

$$\frac{\partial^2 \gamma_{xy}}{\partial x \partial y} = \frac{\partial^2 \varepsilon_x}{\partial y^2} + \frac{\partial^2 \varepsilon_y}{\partial x^2} \tag{5.5.2.1}$$

The internal equilibrium equations are the same as in the case of plane strain. If one carries out the same derivatives as in the previous section, one arrives at the equation

$$\nabla^2 \left(\sigma_x + \sigma_y\right) = -\left(1 + \nu\right) \nabla \left(\mathbf{b} - \rho \mathbf{a}\right) \tag{5.5.2.2}$$

As with plane strain, if the volume loads and the accelerations are constant, the above equation is reduced to

$$\nabla^2 \left(\sigma_x + \sigma_y\right) = 0 \tag{5.5.2.3}$$

In addition to this, the remaining compatibility conditions must be met, i.e.

$$\frac{\partial^2 \varepsilon_z}{\partial y^2} = \frac{\partial^2 \varepsilon_z}{\partial x^2} = \frac{\partial^2 \varepsilon_z}{\partial x \partial y} = 0 \tag{5.5.2.4}$$

Again, the expressions for the stress field can be obtained from the expressions

$$\sigma_x = \frac{\partial^2 \Phi_1}{\partial y^2} \qquad \sigma_y = \frac{\partial^2 \Phi_1}{\partial x^2} \tag{5.5.2.5a}$$

$$\tau_{xy} = -\frac{\partial^2 \Phi_1}{\partial x \partial y} - (b_x - \rho a_x)\, y - \left(b_y - \rho a_y\right) x \tag{5.5.2.5b}$$

The compatibility equation (5.5.2.3) remains the same as in plane strain

$$\nabla^4 \Phi_1 = \frac{\partial^4 \Phi_1}{\partial x^4} + 2\frac{\partial^4 \Phi_1}{\partial x^2 \partial y^2} + \frac{\partial^4 \Phi_1}{\partial y^4} = 0 \qquad (5.5.2.6)$$

One can demonstrate that the function Φ_1 is related to the Airy function Φ of plane strain in the following way

$$\Phi_1 = \Phi - \frac{\nu \psi z^2}{2(1+\nu)} \qquad (5.5.2.7)$$

where $\psi(x, y)$ is a function that complies with

$$\nabla^2 \psi = 0 \qquad (5.5.2.8)$$

In cases that can be defined as plane stress problems because the thickness is very small, one can ignore the last term of (5.5.2.7). Thus the Airy function of plane stress coincides with the plane strain. However, one should keep in mind that the compatibility equations (5.5.2.4) must also be satisfied.

5.6 REPRESENTATIVE POINTS AND LINES IN PLANE ELASTICITY

The name *singular points* is given to those that meet the condition

$$\sigma_x = \sigma_y \quad \tau_{xy} = 0 \qquad (5.6.1)$$

and one can observe that, at these points, all the directions are principal.

The singular points that also satisfy the following conditions are called *neutral points*

$$\sigma_x = \sigma_y = \tau_{xy} = 0 \qquad (5.6.2)$$

In a stress field, the principal directions have a different orientation at each point. The envelope lines of each family of principal directions in the elastic body are called *isostatic lines*. There is one for each family of principal directions σ_1 and σ_2, and at each point those lines are orthogonal.

To obtain the equations for the isostatic lines, according to figure 5.6.1, equation (5.3.6) will be applied, i.e.

$$\frac{dy}{dx} = \tan\alpha \quad \tan(2\alpha) = \frac{2\tau_{xy}}{\sigma_x - \sigma_y} = \frac{2\tan\alpha}{1-\tan^2\alpha} \qquad (5.6.3)$$

resulting in

$$\left(\frac{dy}{dx}\right)^2 + \frac{\sigma_x - \sigma_y}{\tau_{xy}}\left(\frac{dy}{dx}\right) - 1 = 0 \qquad (5.6.4)$$

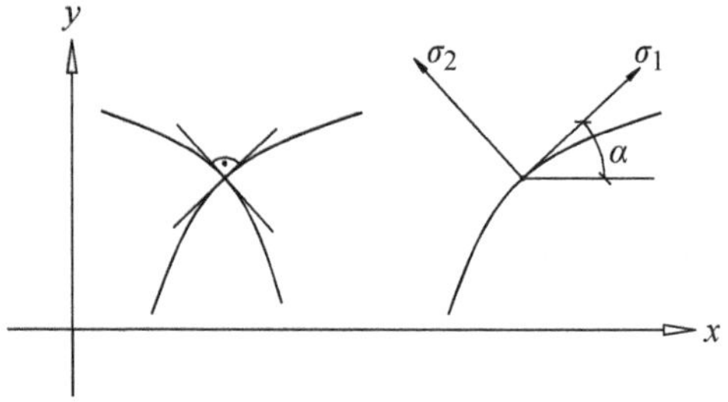

Figure 5.6.1: Isostatic lines

solving the equation, one obtains

$$\frac{dy}{dx} = \frac{\sigma_y - \sigma_x}{2\tau_{xy}} \pm \sqrt{\left(\frac{\sigma_x - \sigma_y}{2\tau_{xy}}\right)^2 + 1} \tag{5.6.5}$$

The two solutions obtained describe the differential equations for each of the families of isostatic lines.

The locus where the orientation of a principal direction is constant is called the *isocline line*. This applies to all the principal directions. Therefore

$$\tan(2\alpha) = \frac{2\tau_{xy}}{\sigma_x - \sigma_y} = \text{constant} \tag{5.6.6}$$

for each inclination of the principal direction; in other words, for each value of α an isocline exists. According to the definition of a singular point, all of the isoclines pass through them.

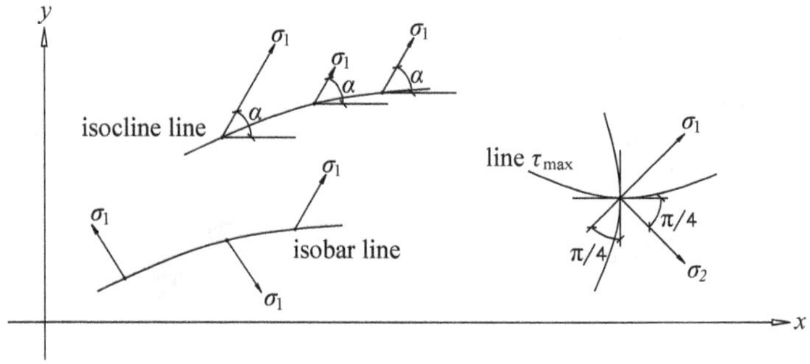

Figure 5.6.2: Lines representing stress fields

The expressions for the principal stresses have been given by (5.3.5). The locus of the points for which σ_1 has the same value is called the *isobar line*, and the same is true for σ_2. As a result, the isobar line equations are

$$\sigma_1 = \frac{\sigma_x + \sigma_y}{2} + \sqrt{\left(\frac{\sigma_x - \sigma_y}{2}\right)^2 + \tau_{xy}^2} = k_1 \tag{5.6.7a}$$

$$\sigma_2 = \frac{\sigma_x + \sigma_y}{2} - \sqrt{\left(\frac{\sigma_x - \sigma_y}{2}\right)^2 + \tau_{xy}^2} = k_2 \tag{5.6.7b}$$

One calls the *maximum shear stress lines*, the locus of the envelopes of the directions of the maximum shear stress. They form two orthogonal families with a 45° inclination with regard to the isostatic lines.

The slope of the maximum shear stress line at a point is $dy/dx = \tan\varphi$ where

$$\varphi = \alpha \pm \frac{\pi}{4} \qquad \tan(2\alpha) = \tan\left(2\varphi \mp \frac{\pi}{2}\right) = -\frac{1}{\tan(2\varphi)} \tag{5.6.8}$$

then

$$\tan(2\alpha) = \frac{2\tau_{xy}}{\sigma_x - \sigma_y} = -\frac{1}{\tan(2\varphi)} = \frac{\tan^2\varphi - 1}{2\tan\varphi} \tag{5.6.9}$$

consequently

$$\left(\frac{dy}{dx}\right)^2 - \frac{4\tau_{xy}}{\sigma_x - \sigma_y}\left(\frac{dy}{dx}\right) - 1 = 0 \tag{5.6.10}$$

and therefore

$$\frac{dy}{dx} = \frac{2\tau_{xy}}{\sigma_x - \sigma_y} \pm \sqrt{\left(\frac{2\tau_{xy}}{\sigma_x - \sigma_y}\right)^2 + 1} \tag{5.6.11}$$

which are the differential equations of the two families of maximum shear stress lines.

5.7 PLANE ELASTICITY IN POLAR COORDINATES

5.7.1 Equilibrium equations and kinematic relationships

On occasions, due to the geometry of the continuum that is studied, it may be convenient to adopt coordinate systems that are not Cartesian. For example, the use of polar coordinates may be useful in cases with symmetry of revolution or semi-infinite bodies, amongst many others. Some of these situations appear in figure 5.7.1.1. The definitions of the stresses and the sign convention adopted are shown in figure 5.7.1.2.

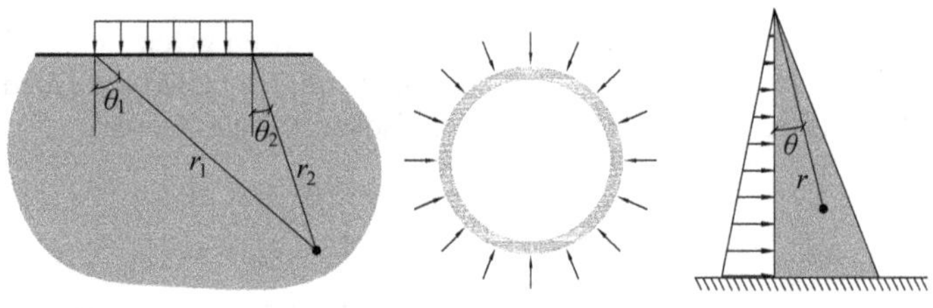

a) Semi-infinite domain b) Hollow tube c) Gravity dam

Figure 5.7.1.1: Examples of continuum media that can be studied in polar coordinates

In the differential element with dimensions dr, $rd\theta$, the equilibrium of forces in the radial direction is written as

$$\left(\sigma_r + \frac{\partial \sigma_r}{\partial r} dr\right)(r + dr)\, d\theta - \sigma_r r d\theta + \left(\tau_{\theta r} + \frac{\partial \tau_{\theta r}}{\partial \theta} d\theta\right) dr - \tau_{\theta r} dr$$
$$- \left(\sigma_\theta + \frac{\partial \sigma_\theta}{\partial \theta} d\theta\right) dr \frac{d\theta}{2} - \sigma_\theta dr \frac{d\theta}{2} + b_r r d\theta dr = 0 \qquad (5.7.1.1a)$$

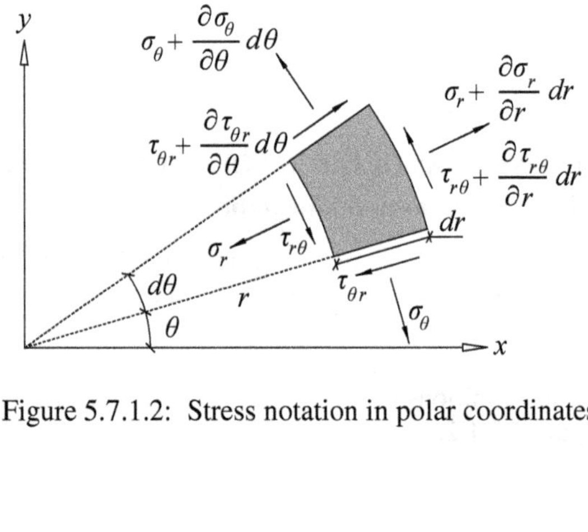

Figure 5.7.1.2: Stress notation in polar coordinates

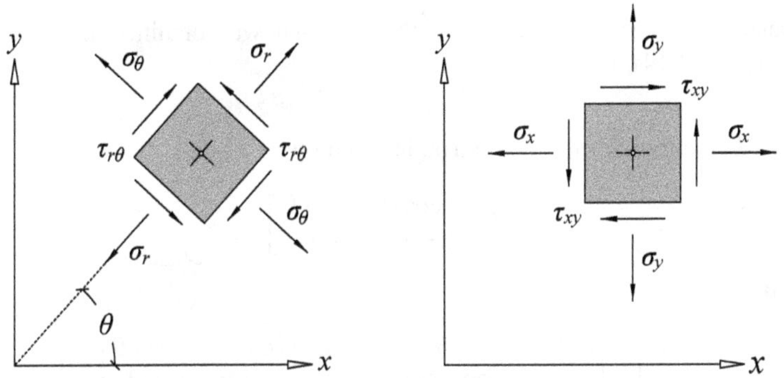

a) Stresses in polar coordinates b) Stresses in Cartesian coordinates

Figure 5.7.1.3: Stresses in polar and Cartesian coordinates

and the forces equilibrium in the circumferential direction is given by

$$\left(\sigma_\theta + \frac{\partial \sigma_\theta}{\partial \theta} d\theta\right) dr - \sigma_\theta dr + \left(\tau_{r\theta} + \frac{\partial \tau_{r\theta}}{\partial r} dr\right)(r + dr)\, d\theta - \tau_{r\theta} r d\theta$$
$$+ \left(\tau_{\theta r} + \frac{\partial \tau_{\theta r}}{\partial \theta} d\theta\right) dr \frac{d\theta}{2} + \tau_{\theta r} dr \frac{dr d\theta}{2} + b_\theta r dr d\theta = 0 \qquad (5.7.1.1b)$$

The equilibrium of moments with respect to z axis, normal to the plane, in the absence of external mass moments is

$$\left(\tau_{r\theta} + \frac{\partial \tau_{r\theta}}{\partial r} dr\right)(r + dr)\, d\theta \frac{dr}{2} + \tau_{r\theta} r d\theta \frac{dr}{2} - \left(\tau_{\theta r} + \frac{\partial \tau_{\theta r}}{\partial \theta} d\theta\right) dr$$
$$\times \left(r + \frac{dr}{2}\right) \frac{d\theta}{2} - \tau_{\theta r} dr \left(r + \frac{dr}{2}\right) \frac{d\theta}{2} = 0 \qquad (5.7.1.1c)$$

simplifying the equations and ignoring infinitesimals of higher order, the following relationships are obtained

$$\sigma_r - \sigma_\theta + r \frac{\partial \sigma_r}{\partial r} + \frac{\partial \tau_{\theta r}}{\partial \theta} + r b_r = 0 \qquad (5.7.1.2a)$$

$$\tau_{\theta r} + \tau_{r\theta} + \frac{\partial \sigma_\theta}{\partial \theta} + r \frac{\partial \tau_{r\theta}}{\partial r} + r b_\theta = 0 \qquad (5.7.1.2b)$$

$$\tau_{r\theta} = \tau_{\theta r} \qquad (5.7.1.2c)$$

Observing figure 5.7.1.3, one concludes that the stress field in the directions of the polar coordinates can be interpreted as a result of giving the Cartesian coordinates a rotation of angle θ. Therefore, the relationship between the corresponding tensors $\boldsymbol{\tau}_{r\theta}, \boldsymbol{\tau}_{xy}$ is the one that was obtained in chapter 3, equation (3.2.14), i.e.

$$\boldsymbol{\tau}_{r\theta} = \mathbf{T}\boldsymbol{\tau}_{xy}\mathbf{T}^T \tag{5.7.1.3}$$

where **T**, the transformation matrix, is given by

$$\mathbf{T} = \begin{bmatrix} \cos\theta & \sin\theta \\ -\sin\theta & \cos\theta \end{bmatrix} \tag{5.7.1.4}$$

and therefore

$$\begin{bmatrix} \sigma_r & \tau_{r\theta} \\ \tau_{r\theta} & \sigma_\theta \end{bmatrix} = \begin{bmatrix} \cos\theta & \sin\theta \\ -\sin\theta & \cos\theta \end{bmatrix} \begin{bmatrix} \sigma_x & \tau_{xy} \\ \tau_{xy} & \sigma_y \end{bmatrix} \begin{bmatrix} \cos\theta & -\sin\theta \\ \sin\theta & \cos\theta \end{bmatrix} \tag{5.7.1.5}$$

Carrying out the appropriate operations, one obtains

$$\sigma_r = \sigma_x \cos^2\theta + \sigma_y \sin^2\theta + \tau_{xy}\sin(2\theta) \tag{5.7.1.6a}$$

$$\sigma_\theta = \sigma_x \sin^2\theta + \sigma_y \cos^2\theta - \tau_{xy}\sin(2\theta) \tag{5.7.1.6b}$$

$$\tau_{r\theta} = -\frac{(\sigma_x - \sigma_y)}{2}\sin(2\theta) + \tau_{xy}\cos(2\theta) \tag{5.7.1.6c}$$

similarly, the inverse relationships can be obtained

$$\sigma_x = \sigma_r \cos^2\theta + \sigma_\theta \sin^2\theta - \tau_{r\theta}\sin(2\theta) \tag{5.7.1.7a}$$

$$\sigma_y = \sigma_r \sin^2\theta + \sigma_\theta \cos^2\theta + \tau_{r\theta}\sin(2\theta) \tag{5.7.1.7b}$$

$$\tau_{xy} = \frac{(\sigma_r - \sigma_\theta)}{2}\sin(2\theta) + \tau_{r\theta}\cos(2\theta) \tag{5.7.1.7c}$$

The relationship between displacements and strains can be obtained from the scheme shown in figure 5.7.1.4, where u, v are the displacements of point O.

The radial ε_r and circumferential ε_θ strains and the angular strain $\gamma_{r\theta}$ result from the following equations

$$\varepsilon_r = \frac{O'A' - OA}{OA} = \frac{\left(dr + u + \dfrac{\partial u}{\partial r}dr - u\right) - dr}{dr} \tag{5.7.1.8a}$$

$$\varepsilon_\theta = \frac{O'B' - OB}{OB} = \frac{\left[v + \dfrac{\partial v}{\partial \theta}d\theta + (r+u)\,d\theta - v\right] - r d\theta}{r d\theta} \tag{5.7.1.8b}$$

$$\gamma_{r\theta} = \frac{\dfrac{\partial v}{\partial r}dr - \dfrac{v}{r}dr}{dr} + \frac{\dfrac{\partial u}{\partial \theta}d\theta}{r d\theta} \tag{5.7.1.8c}$$

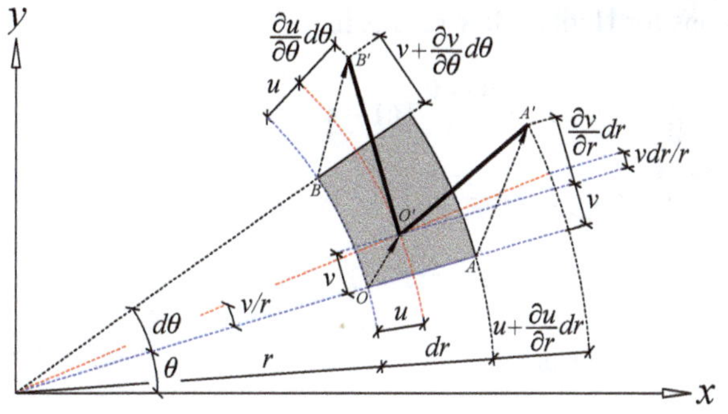

Figure 5.7.1.4: Displacements in a differential element

simplifying the equations, one finally obtains

$$\varepsilon_r = \frac{\partial u}{\partial r} \tag{5.7.1.9a}$$

$$\varepsilon_\theta = \frac{u}{r} + \frac{1}{r}\frac{\partial v}{\partial \theta} \tag{5.7.1.9b}$$

$$\gamma_{r\theta} = \frac{\partial v}{\partial r} + \frac{1}{r}\frac{\partial u}{\partial \theta} - \frac{v}{r} \tag{5.7.1.9c}$$

The only equation of compatibility is obtained by eliminating the displacements u, v from equations (5.7.1.9), giving

$$\frac{\partial^2 \varepsilon_r}{\partial \theta^2} + r\frac{\partial^2 (r\varepsilon_\theta)}{\partial r^2} - r\frac{\partial \varepsilon_r}{\partial r} - r\frac{\partial^2 \gamma_{r\theta}}{\partial r \partial \theta} - \frac{\partial \gamma_{r\theta}}{\partial \theta} = 0 \tag{5.7.1.10}$$

The boundary conditions correspond to the values imposed on the displacements or on the equilibrium of stresses at some points of the boundary.

5.7.2 Plane stress and plane strain fields

The plane strain field in polar coordinates complies with the conditions that

$$\varepsilon_z = 0 \quad \sigma_z = \nu\left(\sigma_r + \sigma_\theta\right) \tag{5.7.2.1}$$

the expressions for Hooke's law results in

$$\varepsilon_r = \frac{1+\nu}{E}\left[(1-\nu)\,\sigma_r - \nu\sigma_\theta\right] \tag{5.7.2.2a}$$

$$\varepsilon_\theta = \frac{1+\nu}{E}\left[(1-\nu)\,\sigma_\theta - \nu\sigma_r\right] \tag{5.7.2.2b}$$

$$\gamma_{r\theta} = \frac{\tau_{r\theta}}{G} \qquad \gamma_{rz} = \gamma_{\theta z} = 0 \tag{5.7.2.2c}$$

Lamé's expressions are

$$\sigma_r = \frac{E}{(1+\nu)(1-2\nu)} \cdot \left[(1-\nu)\,\varepsilon_r - \nu\varepsilon_\theta\right] \tag{5.7.2.3a}$$

$$\sigma_\theta = \frac{E}{(1+\nu)(1-2\nu)} \cdot \left[(1-\nu)\,\varepsilon_\theta - \nu\varepsilon_r\right] \tag{5.7.2.3b}$$

$$\tau_{r\theta} = G\gamma_{r\theta} \qquad \tau_{rz} = \tau_{\theta z} = 0 \tag{5.7.2.3c}$$

In the plane stress field

$$\sigma_z = 0 \quad \varepsilon_z = -\frac{\nu\,(\sigma_r + \sigma_\theta)}{E} \tag{5.7.2.4}$$

the expressions for Hooke's law results in

$$\varepsilon_r = \frac{1}{E}\,(\sigma_r - \nu\sigma_\theta) \tag{5.7.2.5a}$$

$$\varepsilon_\theta = \frac{1}{E}\,(\sigma_\theta - \nu\sigma_r) \tag{5.7.2.5b}$$

$$\gamma_{r\theta} = \frac{\tau_{r\theta}}{G} \qquad \gamma_{rz} = \gamma_{\theta z} = 0 \tag{5.7.2.5c}$$

and Lamé's expressions are

$$\sigma_r = \frac{E}{1-\nu^2}\,(\varepsilon_r + \nu\varepsilon_\theta) \tag{5.7.2.6a}$$

$$\sigma_\theta = \frac{E}{1-\nu^2}\,(\varepsilon_\theta + \nu\varepsilon_r) \tag{5.7.2.6b}$$

$$\tau_{r\theta} = G\gamma_{r\theta} \qquad \tau_{rz} = \tau_{\theta z} = 0 \tag{5.7.2.6c}$$

If in the plane strain field the derivatives that appear in the compatibility equation (5.7.1.10) are carried out on expressions 5.7.2.2 of Hooke's laws, proceeding as one did in section 5.5.1 with the Cartesian coordinates, one finally obtains the expression

$$\nabla^2 (\sigma_r + \sigma_\theta) = \frac{1}{\nu - 1} \nabla \mathbf{b} \tag{5.7.2.7}$$

where $\mathbf{b}$ is the force vector of mass $\mathbf{b}^T = [b_r, b_\theta]$ and where the operator ∇^2 is expressed as

$$\nabla^2 = \frac{\partial^2}{\partial r^2} + \frac{1}{r}\frac{\partial}{\partial r} + \frac{1}{r^2}\frac{\partial^2}{\partial \theta^2} \tag{5.7.2.8}$$

if the mass forces are equal to zero, equation (5.7.2.7) reduces to

$$\nabla^2 (\sigma_r + \sigma_\theta) = 0 \tag{5.7.2.9}$$

5.7.3 Airy function in polar coordinates

As in the case of Cartesian coordinates, one can also define a function Φ from which the stresses can be obtained. In the event that there are no mass forces, the expressions are

$$\sigma_r = \frac{1}{r}\frac{\partial \Phi}{\partial r} + \frac{1}{r^2}\frac{\partial^2 \Phi}{\partial \theta^2} \tag{5.7.3.1a}$$

$$\sigma_\theta = \frac{\partial^2 \Phi}{\partial r^2} \tag{5.7.3.1b}$$

$$\tau_{r\theta} = -\frac{\partial}{\partial r}\left(\frac{1}{r}\frac{\partial \Phi}{\partial \theta}\right) = \frac{1}{r^2}\frac{\partial \Phi}{\partial \theta} - \frac{1}{r}\frac{\partial^2 \Phi}{\partial r \partial \theta} \tag{5.7.3.1c}$$

These expressions automatically satisfy the internal equilibrium equations (5.7.1.2). Also, one can verify that by introducing (5.7.3.1) into equation (5.7.2.9), the latter becomes

$$\nabla^4 \Phi = 0 \tag{5.7.3.2}$$

that is to say

$$\left(\frac{\partial^2}{\partial r^2} + \frac{1}{r}\frac{\partial}{\partial r} + \frac{1}{r^2}\frac{\partial^2}{\partial \theta^2}\right)\left(\frac{\partial^2 \Phi}{\partial r^2} + \frac{1}{r}\frac{\partial \Phi}{\partial r} + \frac{1}{r^2}\frac{\partial^2 \Phi}{\partial \theta^2}\right) = 0 \tag{5.7.3.3}$$

The above equation is that which the Airy function must comply with. The mathematical expression for Φ that is convenient to use varies depending on the problem studied. One can verify this in the following section, where several cases that are of practical interest are described.

5.8 EXAMPLES OF ELASTIC PROBLEMS IN POLAR COORDINATES

5.8.1 Circular tube solicited by radial pressure

This consists of a plane strain problem, which is defined by the tube's geometry, with radius r_1, r_2 and the values of the internal and external pressures being p_1, p_2. From this general case, and based on the values of these parameters, several specific cases can be obtained. Figure 5.8.1.1a represents the general approach and figures 5.8.1.1b to 5.8.1.1d show the cases of a thin-walled tube, a tunnel and a solid cylinder.

The expression for the Airy function that fulfils equation (5.7.2.9) and is suited to these problems is

$$\Phi = A \ln r + Br^2 + C \tag{5.8.1.1}$$

The boundary conditions correspond to stresses on the exterior and interior circumferences

$$r = r_i \quad \sigma_r = -p_i \quad i = 1, 2 \tag{5.8.1.2}$$

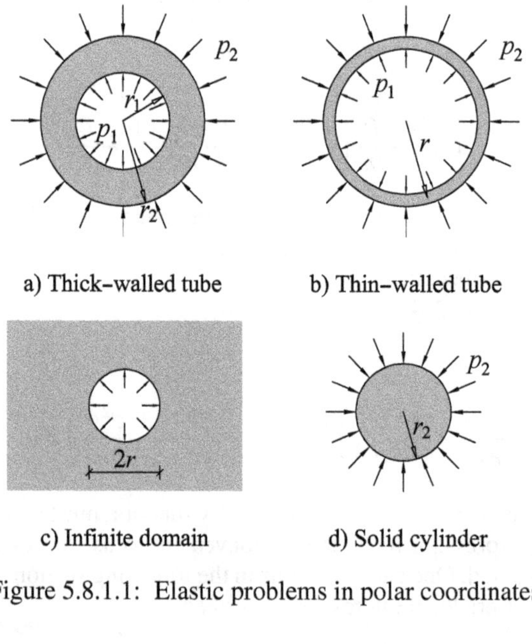

Figure 5.8.1.1: Elastic problems in polar coordinates

Using equations (5.7.3.1), the stress expressions are

$$\sigma_r = \frac{A}{r^2} + 2B \tag{5.8.1.3a}$$

$$\sigma_\theta = -\frac{A}{r^2} + 2B \tag{5.8.1.3b}$$

$$\tau_{r\theta} = 0 \tag{5.8.1.3c}$$

applying the boundary conditions, one obtains

$$A = \frac{r_1^2 r_2^2 (p_2 - p_1)}{r_2^2 - r_1^2} \qquad B = \frac{r_1^2 p_1 - r_2^2 p_2}{2\left(r_2^2 - r_1^2\right)} \tag{5.8.1.4}$$

and substituting in (5.8.1.3) gives

$$\sigma_r = \frac{1}{r_2^2 - r_1^2}\left[r_1^2 p_1 - r_2^2 p_2 + \frac{r_1^2 r_2^2}{r^2}(p_2 - p_1)\right] \tag{5.8.1.5a}$$

$$\sigma_\theta = \frac{1}{r_2^2 - r_1^2}\left[r_1^2 p_1 - r_2^2 p_2 - \frac{r_1^2 r_2^2}{r^2}(p_2 - p_1)\right] \tag{5.8.1.5b}$$

$$\tau_{r\theta} = 0 \tag{5.8.1.5c}$$

Due to the symmetry of the problem, the circumferential displacement is equal to zero. To obtain the radial displacement, one has to use the expressions for Hooke's law, equations (5.7.2.2)

$$\frac{du}{dr} = \varepsilon_r = \frac{1+\nu}{E}\left[(1-\nu)\,\sigma_r - \nu\sigma_\theta\right] \tag{5.8.1.6a}$$

$$\frac{u}{r} = \varepsilon_\theta = \frac{1+\nu}{E}\left[(1-\nu)\,\sigma_\theta - \nu\sigma_r\right] \tag{5.8.1.6b}$$

by using any of equations (5.8.1.6), one arrives at the following expression

$$u = \frac{1+\nu}{E\left(r_2^2 - r_1^2\right)}\left[(1-2\nu)\left(r_1^2 p_1 - r_2^2 p_2\right) r - \frac{r_1^2 r_2^2}{r}(p_2 - p_1)\right] \tag{5.8.1.7}$$

If only pressure p_1 or p_2 exists, as shown in figure 5.8.1.2, the previous expressions are simplified.

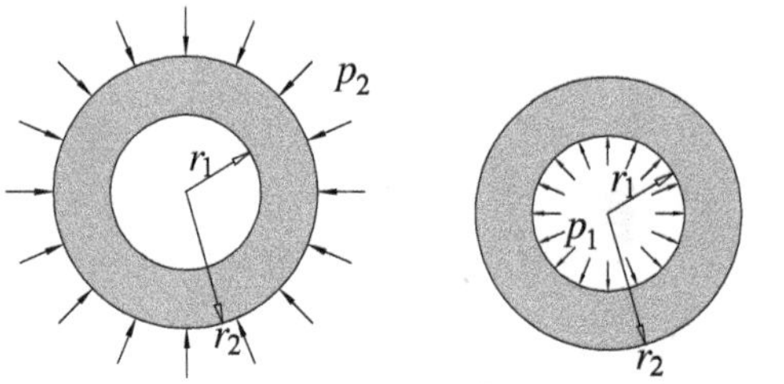

a) Tube under external pressure b) Tube under internal pressure

Figure 5.8.1.2: Tube under external or internal pressure

When there is only external pressure, the stresses are

$$\sigma_r = -\frac{r_2^2 p_2}{r_2^2 - r_1^2}\left(1 - \frac{r_1^2}{r^2}\right) \tag{5.8.1.8a}$$

$$\sigma_\theta = -\frac{r_2^2 p_2}{r_2^2 - r_1^2}\left(1 + \frac{r_1^2}{r^2}\right) \tag{5.8.1.8b}$$

$$\tau_{r\theta} = 0 \tag{5.8.1.8c}$$

and the expression for the radial displacement becomes

$$u = -\frac{(1+\nu)\, r_2^2 p_2}{E\left(r_2^2 - r_1^2\right) r}\left[(1-2\nu)\, r^2 + r_1^2\right] \tag{5.8.1.9}$$

Similarly, when there is only internal pressure, one obtains the following expressions for the stresses and displacement u

$$\sigma_r = \frac{r_1^2 p_1}{r_2^2 - r_1^2}\left(1 - \frac{r_2^2}{r^2}\right) \tag{5.8.1.10a}$$

$$\sigma_\theta = \frac{r_1^2 p_1}{r_2^2 - r_1^2}\left(1 + \frac{r_2^2}{r^2}\right) \tag{5.8.1.10b}$$

$$\tau_{r\theta} = 0 \tag{5.8.1.10c}$$

and

$$u = \frac{(1+\nu)\, r_1^2 p_1}{E\left(r_2^2 - r_1^2\right) r}\left[(1-2\nu)\, r^2 + r_2^2\right] \tag{5.8.1.11}$$

The solid cylinder corresponds to the situation in which $p_1 = 0$ and $r_1 = 0$; thus the expressions reduce to

$$\sigma_r = \sigma_\theta = -p_2 \tag{5.8.1.12}$$

$$u = -\frac{(1+\nu)(1-2\nu)\, r p_2}{E} \tag{5.8.1.13}$$

Thin-walled tubes with thickness t can be studied assuming the following simplifications

$$r = \frac{r_1 + r_2}{2} \tag{5.8.1.14a}$$

$$r_1^2 \simeq r_2^2 \simeq r^2 \tag{5.8.1.14b}$$

$$r_2^2 - r_1^2 = (r_2 - r_1)(r_2 + r_1) = 2rt \tag{5.8.1.14c}$$

the stress expressions and the radial displacement become

$$\sigma_r = 0 \tag{5.8.1.15a}$$

$$\sigma_\theta = \frac{(p_1 - p_2)\, r}{t} \tag{5.8.1.15b}$$

$$\tau_{r\theta} = 0 \tag{5.8.1.15c}$$

and

$$u = \frac{(1-\nu^2)(p_1 - p_2)\, r^2}{Et} \tag{5.8.1.16}$$

A continuum with a hole, which can be interpreted as a tunnel, corresponds to $p_2 = 0,\ r_2 = \infty$. Therefore, the stresses and the radial displacement become

$$\sigma_r = -\sigma_\theta = -\frac{r_1^2}{r^2} p_1 \tag{5.8.1.17}$$

$$u = \frac{(1+\nu)\, r_1^2 p_1}{Er} \tag{5.8.1.18}$$

In this particular case, it is also interesting to know the stress tensor in Cartesian axes. Applying equations (5.7.1.7), one obtains

$$\sigma_x = -\frac{r_1^2 p_1 \left(\cos^2\theta - \sin^2\theta\right)}{r^2} = -\frac{r_1^2 p_1 \left(x^2 - y^2\right)}{\left(x^2 + y^2\right)^2} \qquad (5.8.1.19a)$$

$$\sigma_y = -\frac{r_1^2 p_1 \left(\sin^2\theta - \cos^2\theta\right)}{r^2} = -\frac{r_1^2 p_1 \left(y^2 - x^2\right)}{\left(x^2 + y^2\right)^2} \qquad (5.8.1.19b)$$

$$\tau_{xy} = -\frac{r_1^2 p_1}{r^2}\sin(2\theta) = -\frac{2r_1^2 p_1 xy}{\left(x^2 + y^2\right)^2} \qquad (5.8.1.19c)$$

5.8.2 Wedge and semi-infinite continuum with a concentrated vertical load

Suppose there exists a plane strain field in a space defined by a wedge with angle 2α and a load N in its axis of symmetry, as shown in figure 5.8.2.1. In this case, it is advisable to define the Airy function using the following expression

$$\Phi = Ar\theta \sin\theta \qquad (5.8.2.1)$$

The stresses become

$$\sigma_r = 2A\frac{\cos\theta}{r} \qquad (5.8.2.2a)$$

$$\sigma_\theta = \tau_{r\theta} = 0 \qquad (5.8.2.2b)$$

The value of constant A can be obtained by establishing, along any curve with constant radius, the equilibrium between force N and the integral of the

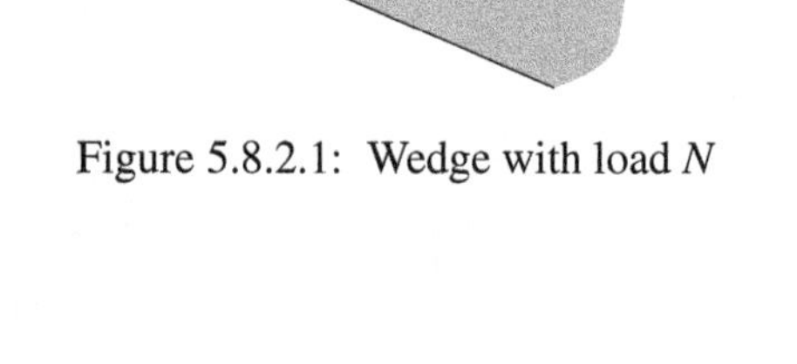

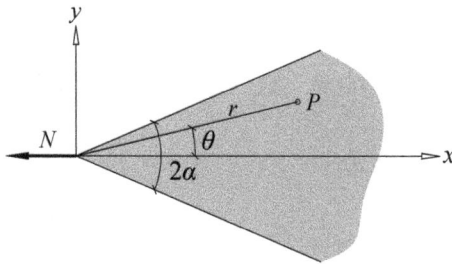

Figure 5.8.2.1: Wedge with load N

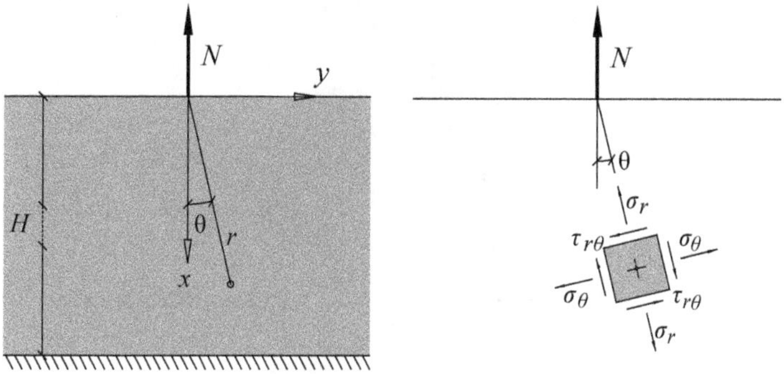

Figure 5.8.2.2: Semi-infinite body with concentrated load N

horizontal component of stress σ_r, i.e.

$$N = \int_{-\alpha}^{\alpha} \sigma_r \cos\theta r d\theta \tag{5.8.2.3}$$

substituting for the expression σ_r and carrying out the integration gives

$$A = \frac{N}{2\alpha + \sin(2\alpha)} \tag{5.8.2.4}$$

thus, the last stress expression is

$$\sigma_r = \frac{2N\cos\theta}{[2\alpha + \sin(2\alpha)]\, r} \tag{5.8.2.5a}$$

$$\sigma_\theta = \tau_{r\theta} = 0 \tag{5.8.2.5b}$$

A particular case of this problem is when $\alpha = \pi/2$, which corresponds to a semi-infinite domain under a concentrated load, which is a soil mechanics application. Generally, this continuum will have at a depth H a layer of material that can be considered undeformable, as shown in figure 5.8.2.2.

The stress expressions are obtained from (5.8.2.5) and become

$$\sigma_r = \frac{2N\cos\theta}{\pi r} \tag{5.8.2.6a}$$

$$\sigma_\theta = \tau_{r\theta} = 0 \tag{5.8.2.6b}$$

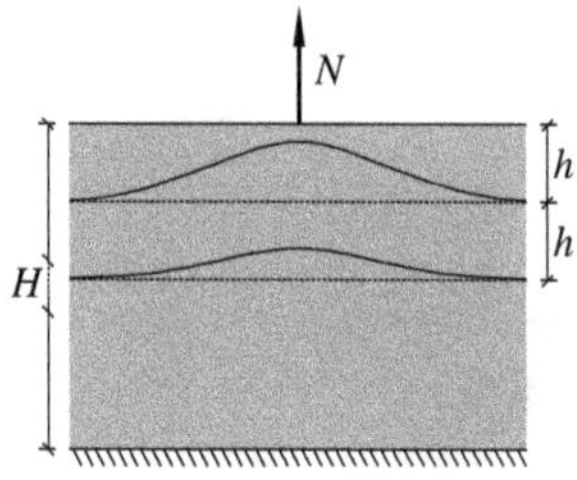

Figure 5.8.2.3: Distribution of σ_r in a semi-infinite domain

Applying equations (5.7.1.7) to this case, one obtains the components of the stress tensor in Cartesian coordinates

$$\sigma_x = \frac{2N\cos^3\theta}{\pi r} = \frac{2Nx^3}{\pi\left(x^2+y^2\right)^2} \tag{5.8.2.7a}$$

$$\sigma_y = \frac{2N\sin^2\theta\cos\theta}{\pi r} = \frac{2Ny^2x}{\pi\left(x^2+y^2\right)^2} \tag{5.8.2.7b}$$

$$\tau_{xy} = \frac{2N\cos^2\theta\sin\theta}{\pi r} = \frac{2Nx^2y}{\pi\left(x^2+y^2\right)^2} \tag{5.8.2.7c}$$

At the point on which the load is applied, the stress has infinite value, which makes no physical sense and is due to the definition of the load as a concentrated force N. In figure 5.8.2.3, one can observe the distribution of stress σ_r at different depths.

To obtain the displacements, one uses the kinematic relationships (5.7.1.9) and the equations for Hooke's law (5.7.2.2) and substitutes the expressions of stress from (5.8.2.6). Following this, one arrives at

$$\varepsilon_r = \frac{\partial u}{\partial r} = \frac{2\left(1-\nu^2\right)N\cos\theta}{\pi E r} \tag{5.8.2.8a}$$

$$\varepsilon_\theta = \frac{u}{r} + \frac{1}{r}\frac{\partial v}{\partial\theta} = -\frac{2\nu\left(1+\nu\right)N\cos\theta}{\pi E r} \tag{5.8.2.8b}$$

$$\gamma_{r\theta} = \frac{1}{r}\frac{\partial u}{\partial\theta} + \frac{\partial v}{\partial r} - \frac{v}{r} = 0 \tag{5.8.2.8c}$$

integrating the first equation gives

$$u = \frac{2\left(1-\nu^2\right)N\cos\theta}{\pi E}\ln r + f_1\left(\theta\right) \tag{5.8.2.9}$$

substituting into the second equation and integrating again, one obtains

$$v = -\frac{2\nu(1+\nu)N\sin\theta}{\pi E} - \frac{2(1-\nu^2)N\sin\theta}{\pi E}\ln r - \int f_1(\theta)\,d\theta + f_2(r) \tag{5.8.2.10}$$

introducing u, v in the third equation and carrying out the corresponding integrals, one obtains the following expressions

$$u = \frac{2(1-\nu^2)N\cos\theta}{\pi E}\ln r + \frac{(1+\nu)(1-2\nu)N\theta\sin\theta}{\pi E} + B\sin\theta + C\cos\theta \tag{5.8.2.11a}$$

substituting into the second equation and integrating gives

$$v = -\frac{2\nu(1+\nu)N\sin\theta}{\pi E} - \frac{2(1-\nu^2)N\sin\theta\ln r}{\pi E} + \frac{(1+\nu)(1-2\nu)N\theta\cos\theta}{\pi E} - \frac{(1+\nu)(1-2\nu)N\sin\theta}{\pi E} + B\cos\theta - C\sin\theta + Dr \tag{5.8.2.11b}$$

The following boundary conditions can be applied due to the symmetry

when $\theta = 0 \Rightarrow v = 0$
when $r = H$ y $\theta = 0 \Rightarrow u = 0$

which allows one to conclude that

$$B = D = 0 \tag{5.8.2.12a}$$

$$C = -\frac{2(1-\nu^2)N\ln H}{\pi E} \tag{5.8.2.12b}$$

therefore, the final expressions for displacements are

$$u = \frac{2(1-\nu^2)N}{\pi E}\left[\ln\left(\frac{r}{H}\right)\cos\theta + \frac{(1-2\nu)\,\theta\sin\theta}{2(1-\nu)}\right] \tag{5.8.2.13a}$$

$$v = -\frac{2(1-\nu^2)N}{\pi E}\left\{\left[\ln\left(\frac{r}{H}\right) + \frac{1}{2(1-\nu)}\right]\sin\theta - \frac{(1-2\nu)\,\theta\cos\theta}{2(1-\nu)}\right\} \tag{5.8.2.13b}$$

The equations of the isostatic lines can be obtained from equation (5.6.5)

$$\frac{dy}{dx} = \frac{\sigma_y - \sigma_x}{2\tau_{xy}} \pm \sqrt{\left(\frac{\sigma_x - \sigma_y}{2\tau_{xy}}\right)^2 + 1} \tag{5.8.2.14}$$

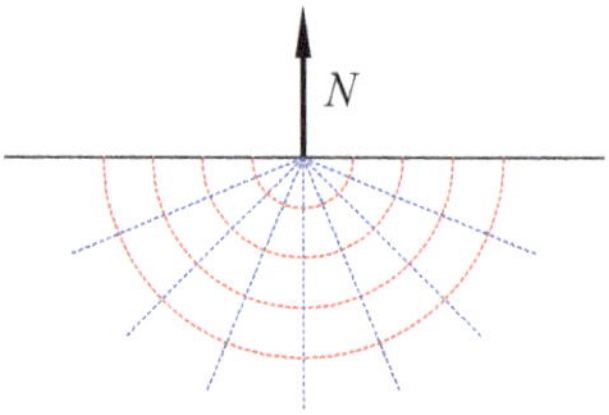

Figure 5.8.2.4: Geometry of the isostatic lines of the semi-infinite domain

if one substitutes the stresses obtained in (5.8.2.7), it turns out that

$$\frac{dy}{dx} = -\frac{\cos^2\theta - \sin^2\theta}{2\sin\theta\cos\theta} \pm \frac{1}{2\sin\theta\cos\theta} \tag{5.8.2.15}$$

operating on this expression, one can verify that the two families of isostatic lines are

$$\frac{dy_1}{dx} = \frac{\sin\theta}{\cos\theta} \quad \frac{dy_2}{dx} = -\frac{\cos\theta}{\sin\theta} \tag{5.8.2.16}$$

observing figure 5.8.2.2, one can verify that

$$\cos\theta = \frac{x}{\sqrt{x^2+y^2}} \qquad \sin\theta = \frac{y}{\sqrt{x^2+y^2}} \tag{5.8.2.17}$$

and substituting into (5.8.2.16) gives

$$\frac{dy_1}{dx} = \frac{y}{x} \quad \frac{dy_2}{dx} = -\frac{x}{y} \tag{5.8.2.18}$$

after performing the integrations, one obtains

$$y_1 = k_1 \cdot x \quad y_2^2 + x^2 = k_2 \tag{5.8.2.19}$$

The first of the families corresponds to the straight lines that pass through the coordinates origin, this being the point where the load is applied, and the second family are circles whose centre is also on that point. In figure 5.8.2.4, some of those isostatic lines are drawn in.

5.8.3 Semi-infinite domain with a distributed load

The previous example can be extended to the case where a distributed load exists on a segment of the semi-infinite domain's surface, as shown in figure 5.8.3.1.

From figure 5.8.3.1b, one observes that

$$p(y)\,dy\cos\theta = p(y)\,rd\theta \qquad (5.8.3.1)$$

using expressions (5.8.2.7) from the preceding section, the stresses can be defined as

$$\sigma_x = -\frac{2}{\pi}\int_{\theta_1}^{\theta_2}\frac{p(y)\,r\cos^3\theta}{r\cos\theta}d\theta = -\frac{2}{\pi}\int_{\theta_1}^{\theta_2}p(y)\cos^2\theta d\theta \qquad (5.8.3.2a)$$

$$\sigma_y = -\frac{2}{\pi}\int_{\theta_1}^{\theta_2}\frac{p(y)\,r\sin^2\theta\cos\theta}{r\cos\theta}d\theta = -\frac{2}{\pi}\int_{\theta_1}^{\theta_2}p(y)\sin^2\theta d\theta \qquad (5.8.3.2b)$$

$$\tau_{xy} = -\frac{2}{\pi}\int_{\theta_1}^{\theta_2}\frac{p(y)\,r\cos^2\theta\sin\theta}{r\cos\theta}d\theta = -\frac{2}{\pi}\int_{\theta_1}^{\theta_2}p(y)\sin\theta\cos\theta d\theta$$

(5.8.3.2c)

In the case that the distributed load is uniform, these integrals give the following results

$$\sigma_x = -\frac{p}{2\pi}\left[2(\theta_2-\theta_1)+\sin(2\theta_2)-\sin(2\theta_1)\right] \qquad (5.8.3.3a)$$

$$\sigma_y = -\frac{p}{2\pi}\left[2(\theta_2-\theta_1)-\sin(2\theta_2)+\sin(2\theta_1)\right] \qquad (5.8.3.3b)$$

$$\tau_{xy} = \frac{p}{2\pi}\left[\cos(2\theta_2)-\cos(2\theta_1)\right] \qquad (5.8.3.3c)$$

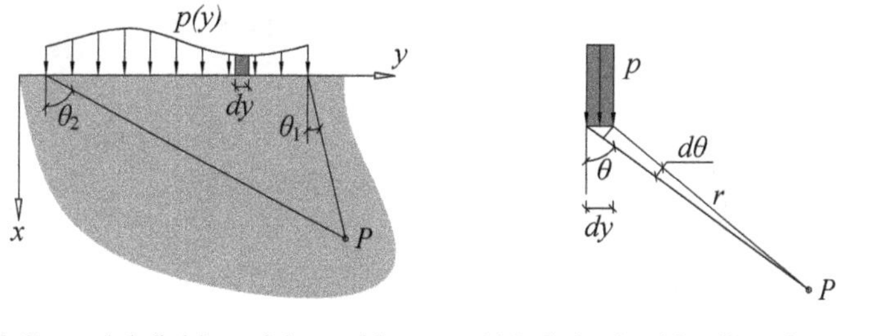

a) General definition of the problem b) Infinitesimal loading element

Figure 5.8.3.1: Semi-infinite domain with a distributed load

The stress expressions in polar coordinates can be obtained from the relationship (5.7.1.6), i.e.

$$\sigma_r = -\frac{p}{2\pi}\left[2\left(\theta_2-\theta_1\right)+\sin\left(2\theta_2+2\theta\right)-\sin\left(2\theta_1+2\theta\right)\right] \tag{5.8.3.4a}$$

$$\sigma_\theta = -\frac{p}{2\pi}\left[2\left(\theta_2-\theta_1\right)-\sin\left(2\theta_2+2\theta\right)+\sin\left(2\theta_1+2\theta\right)\right] \tag{5.8.3.4b}$$

$$\tau_{r\theta} = \frac{p}{2\pi}\left[\cos\left(2\theta_2+2\theta\right)-\cos\left(2\theta_1+2\theta\right)\right] \tag{5.8.3.4c}$$

The principal stresses are given by equation (3.4.8) and results in

$$\sigma_1 = -\frac{p}{\pi}\left[\theta_2-\theta_1+\sin\left(\theta_2-\theta_1\right)\right] \tag{5.8.3.5a}$$

$$\sigma_2 = -\frac{p}{\pi}\left[\theta_2-\theta_1-\sin\left(\theta_2-\theta_1\right)\right] \tag{5.8.3.5b}$$

5.8.4 Thin plate with hole

Figure 5.8.4.1 shows a thin plate loaded by a distribution of uniform stresses in two parallel boundaries and an interior hole of small dimensions in comparison with those of the plate. Away from the hole, the stress distribution is uniform, but in its vicinity the situation changes; the stress value will increase locally and, as a result, the material can develop a crack. This is a case that has been extensively studied in fracture mechanics and can be analysed as a plane stress problem.

Away from the hole, the stress distribution is uniform and has the value

$$\sigma_x = \sigma \qquad \sigma_y = \tau_{xy} = 0 \tag{5.8.4.1}$$

To obtain the stress distribution near the hole, it is more convenient to formulate the problem in polar coordinates. The following expression is chosen for the Airy function

$$\Phi = A\ln r + Br^2 + \left(Cr^2 + \frac{D}{r^2} + E\right)\cos\left(2\theta\right) \tag{5.8.4.2}$$

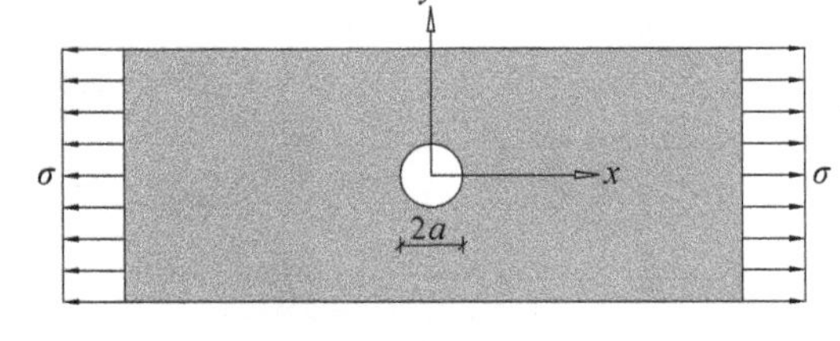

Figure 5.8.4.1: Thin plate with hole

from this, the expressions for the stress are

$$\sigma_r = \frac{A}{r^2} + 2B - \left(2C + \frac{6D}{r^4} + \frac{4E}{r^2}\right)\cos(2\theta) \tag{5.8.4.3a}$$

$$\sigma_\theta = -\frac{A}{r^2} + 2B + \left(2C + \frac{6D}{r^4}\right)\cos(2\theta) \tag{5.8.4.3b}$$

$$\tau_{r\theta} = \left(2C - \frac{6D}{r^4} - \frac{2E}{r^2}\right)\sin(2\theta) \tag{5.8.4.3c}$$

To obtain the values of the coefficients of the Airy function, one has to impose the boundary conditions. For those points located away from the hole, that is to say, for large values of r, the stresses should be those of equation (5.8.4.1). Thus, if one calculates σ_x, σ_y, τ_{xy} using expression (5.7.1.7) and assumes that $r \to \infty$, the values of τ_{xy} are zero and, from the other two equations, one obtains

$$B = \frac{\sigma}{4} \quad C = -\frac{\sigma}{4} \tag{5.8.4.4}$$

In the boundary of the hole there are no applied forces; therefore, at $r = a$ the following condition applies

$$\sigma_r = \tau_{r\theta} = 0 \tag{5.8.4.5}$$

and therefore

$$\frac{A}{a^2} + \frac{\sigma}{2} - \left(-\frac{\sigma}{2} + \frac{6D}{a^4} + \frac{4E}{a^2}\right)\cos(2\theta) = 0 \tag{5.8.4.6a}$$

$$\left(\frac{\sigma}{2} + \frac{6D}{a^4} + \frac{2E}{a^2}\right)\sin(2\theta) = 0 \tag{5.8.4.6b}$$

For these equations to be valid at any point of the boundary of the hole, it is necessary that

$$\frac{A}{a^2} + \frac{\sigma}{2} = 0 \tag{5.8.4.7a}$$

$$-\frac{\sigma}{2} + \frac{6D}{a^4} + \frac{4E}{a^2} = 0 \tag{5.8.4.7b}$$

$$\frac{\sigma}{2} + \frac{6D}{a^4} + \frac{2E}{a^2} = 0 \tag{5.8.4.7c}$$

from which, one obtains

$$A = -\frac{a^2\sigma}{2} \quad D = -\frac{a^4\sigma}{4} \quad E = \frac{a^2\sigma}{2} \tag{5.8.4.8}$$

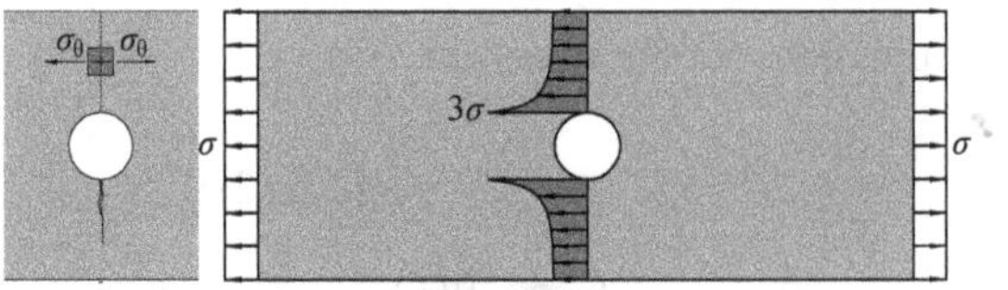

Figure 5.8.4.2: Stress σ_θ along the vertical axis

Table 5.8.4.1: Values of σ_θ for $\theta = \pm\pi/2$

r	a	$2a$	$3a$	$4a$	$5a$
$\sigma_\theta = \sigma_x$	3σ	1.22σ	1.07σ	1.03σ	1.02σ

therefore, the general stress expression becomes

$$\sigma_r = \left[1 - \frac{a^2}{r^2} + \left(1 + \frac{3a^4}{r^4} - \frac{4a^2}{r^2}\right)\cos(2\theta)\right]\frac{\sigma}{2} \tag{5.8.4.9a}$$

$$\sigma_\theta = \left[1 + \frac{a^2}{r^2} - \left(1 + \frac{3a^4}{r^4}\right)\cos(2\theta)\right]\frac{\sigma}{2} \tag{5.8.4.9b}$$

$$\tau_{r\theta} = -\left[\left(1 - \frac{3a^4}{r^4} + \frac{2a^2}{r^2}\right)\sin(2\theta)\right]\frac{\sigma}{2} \tag{5.8.4.9c}$$

Observing (5.8.4.9b), one can verify that the stress σ_θ has a maximum value when $\theta = \pm\pi/2$, that is to say, it coincides with σ_x and results in

$$\sigma_\theta = \sigma_x = \left(2 + \frac{a^2}{r^2} + \frac{3a^4}{r^4}\right)\frac{\sigma}{2} \tag{5.8.4.10}$$

In the top and bottom points of the hole, the stress is $\sigma_\theta = \sigma_x = 3\sigma$, which it is much higher than at any other point on the plate and decreases quickly when moving away from those positions until it reaches asymptotically the value $\sigma_\theta = \sigma$, as can be seen in figure 5.8.4.2. Table 5.8.4.1 illustrates the distribution of these stress values.

The local increase in σ_θ can cause elastic collapse of the material, as well as the appearance of cracks, as shown in figure 5.8.4.2.

EXERCISES

5.1. In the body shown in the figure, the total stress on *AB* plane is horizontal and has a value of 10 MPa. Knowing that $\sigma_1 = 3\sigma_2$, draw Mohr's circle of the principal stresses, the stress ellipse and the normal and shear stresses, acting on BC plane.

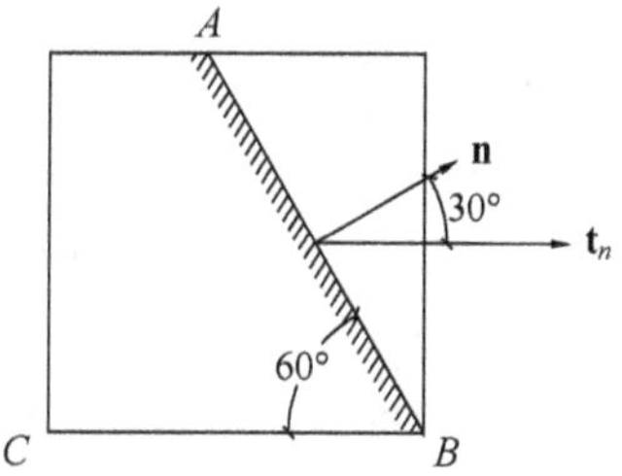

5.2. A rectangular plate with vertices A, B, C, D, whose coordinates are indicated in the table, has its boundary subjected to stresses contained in its plane and varying linearly, where the values in the vertices in MPa are:

	$A\,(-8, 2)$	$B\,(4, 2)$	$C\,(4, -10)$	$D\,(-8, -10)$
σ_x	0	0	0	0
σ_y	62	−22	−58	26
τ_{xy}	24	−12	−12	24

Find:

1. Airy function and stress tensor.
2. Distribution of normal and shear stresses in the AC diagonal.
3. Isoclines that correspond to angles 15°, 30°, 45°, 60° and 80°.
4. Singular points.
5. Equation of the isostatics that are straight lines.

5.3. The vertices of a rectangular plate are, in m:

$$A\,(4,2) \quad B\,(4,-2) \quad C\,(-4,-2) \quad D\,(-4,2)$$

This plate is subjected to a linear distribution of normal stresses on its boundary, whose values in MPa in the vertices are those indicated in the figure below. The shear stress has a parabolic distribution in the vertical sides.

Calculate the following:

1. The expression for the shear stresses and the Airy function.
2. Expressions for σ_x, σ_y and τ_{xy} in a generic point of the plate, and the singular points.
3. Principal stresses at the point (0, 0). Draw the stress ellipse at that point.

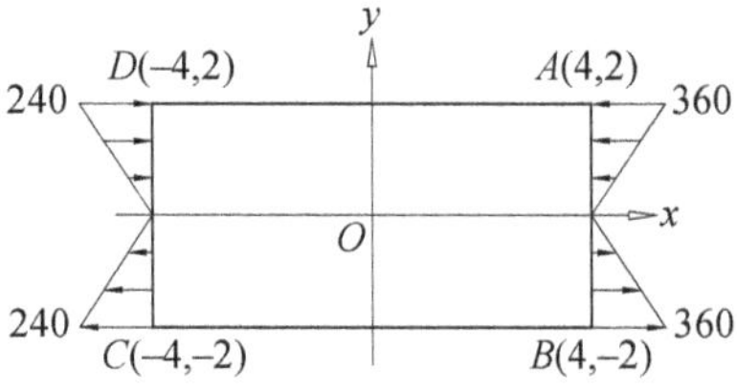

Distribution of normal stresses

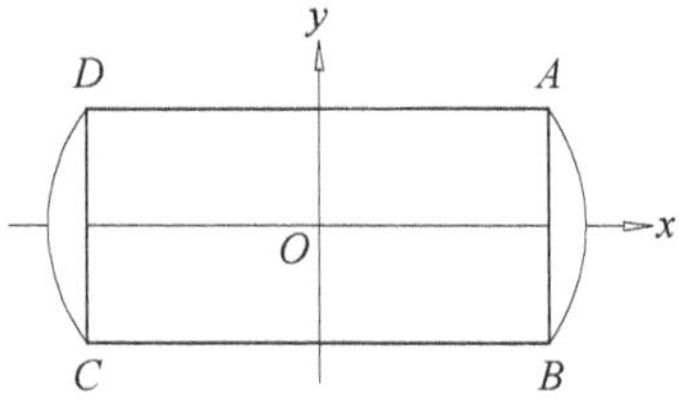

Distribution of shear stresses

5.4. A rectangular plate of vertices A (4,2), B (4,−2), C (−4,−2), D (−4,2) and 1 cm thick is subjected to the following stresses on its boundary.

Side	AB	BC	CD	DA
σ_x (MPa)	−64	–	64	–
σ_y (MPa)	–	$2x^3 - 12x$	–	$2x^3 - 12x$
τ_{xy} (MPa)	$48y$	$-6x^2$	$48y$	$6x^2$

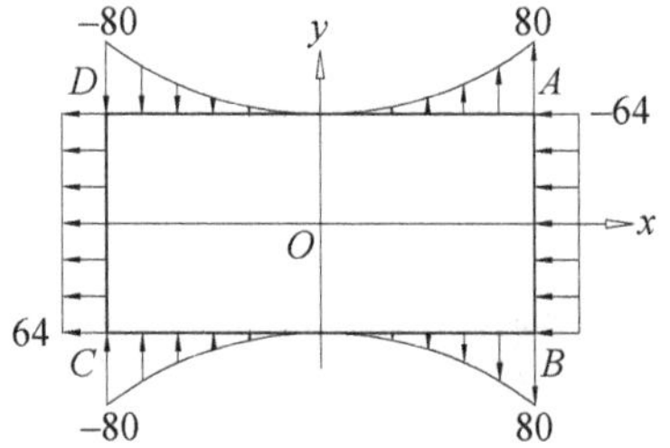

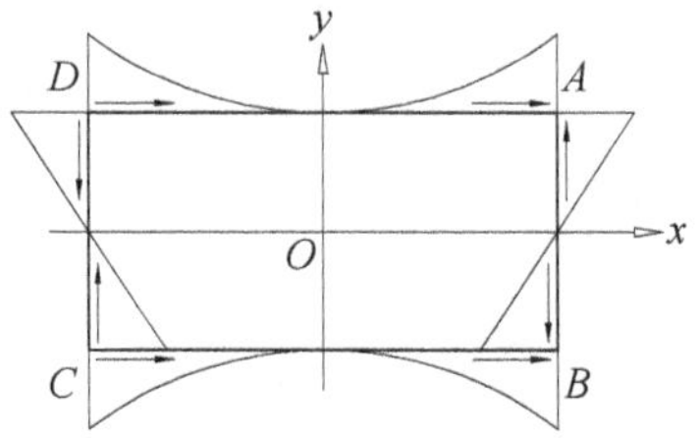

Find:

1. Airy function and stress expression at any point on the plate.
2. Singular points.
3. Isostatic equations, indicating those that are straight.
4. The value of the angles on the vertices after deformation with a modulus $G = 10^3$ MPa.

5.5. The vertices of a rectangular plate have the following coordinates (in m):

$$A(6,4) \quad B(6,-4) \quad C(-6,-4) \quad D(-6,4)$$

The plate is subjected on its boundary to normal stresses that vary linearly and whose values in the vertices are those that appear in the figure (in MPa). The shear stresses are equal to zero all along the boundary.

1. Obtain the Airy function.
2. Identify the singular points, if any.
3. Draw the network of isostatic lines.

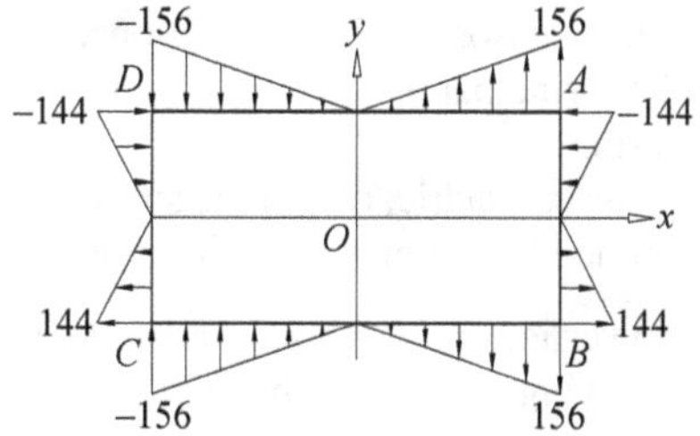

5.6. In the following figures, the stresses distributed on the boundary of a plate are represented. Find:

1. The value of the parabolic distribution of stress σ_x.
2. The expression for the Airy function.
3. Coordinates of the singular points, if any.
4. Principal stresses and directions at the coordinate's origin.

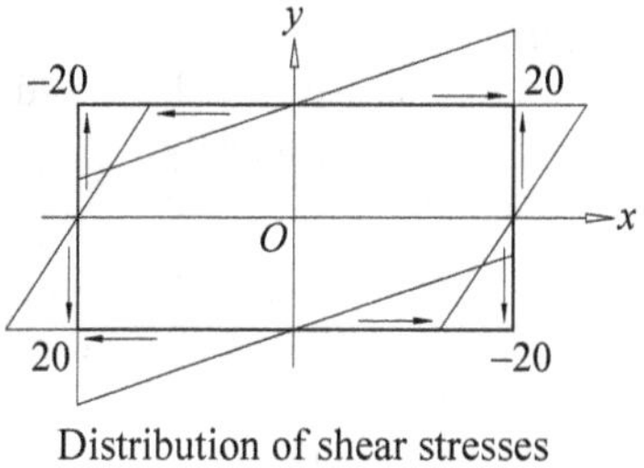

Distribution of shear stresses

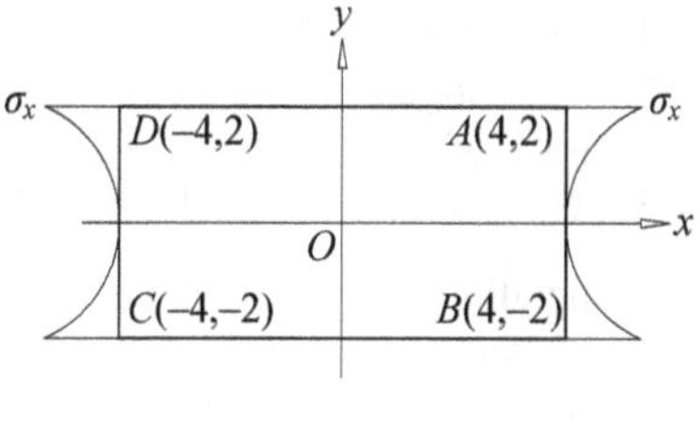

Distribution of normal stresses σ_x

5.7. Consider an elastic, linear and isotropic material with the properties $E = 2000 \cdot a$ and $\nu = 1/3$ in a plane strain field subjected to the stresses shown in the figure.

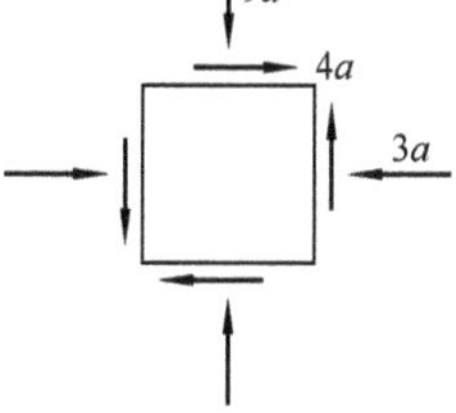

1. Obtain the stress tensor, the hydrostatic stress and the deviatoric tensor.
2. Calculate the principal stresses and directions analytically, and graphically using Mohr's circle. Calculate also the hydrostatic stress and the deviatoric tensor for these directions.
3. Determine the principal strains and their directions. Obtain the volumetric strain.
4. Using the indicatrix quadric, obtain the strain vector in a plane whose direction is defined by the principal direction 1.

5.8. A square plate of vertices *A*, *B*, *C*, *D*, whose coordinates are shown in metres in the following table, is subjected on its boundary to several stresses contained in the same plane and that vary linearly, whose values in the vertices (in MPa) as given in the table below.

	A (0,0)	*B* (0,1)	*C* (1,1)	*D* (1,0)
σ_x	0	−60	−120	−60
σ_y	0	−60	−120	−60
τ_{xy}	0	60	120	60

Find:

1. The stress field at any point of the plate.
2. The singular points, the expressions for the isostatic lines and the lines of maximum shear stress (draw them schematically).
3. The displacement field, knowing that the point *A* is fixed.

5.9. A metallic plate of an elastic, linear and isotropic material is subjected to a plane stress field, as indicated in the figure. The deformations in the directions of the x and y axes are, respectively, $-12.5 \cdot 10^{-5}$ and $-1.136 \cdot 10^{-5}$. Consider $E = 2.2 \cdot 10^5$ MPa and $\nu = 0.25$.

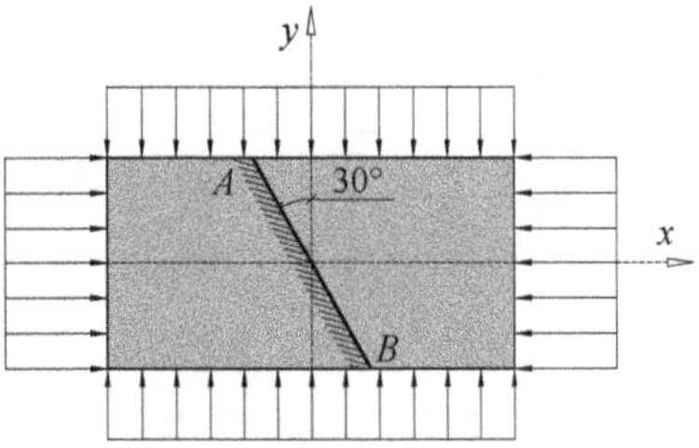

1. Obtain the strain tensor, the stress tensor and the deviatoric tensor.
2. Calculate the stresses, principal directions and invariants.
3. Calculate the stresses on the *AB* plane shown in the figure.
4. Find the value of the maximum shear stress, and the directions of the planes associated with it.
5. Draw the two-dimensional Mohr's circle, indicate the position of the pole, and graphically solve questions 2 and 3.

5.10. A rectangular plate, with the dimensions 40 × 20 cm, is subjected on its boundary to the stresses that are contained in the same plane, as shown in the figures (in MPa). The normal stresses are parabolas of the second degree.

1. Find the stress field of any point of the plate.
2. Find the singular points.
3. Calculate the isostatic network and graphically represent them, distinguishing the two families using continuous and broken lines, respectively.

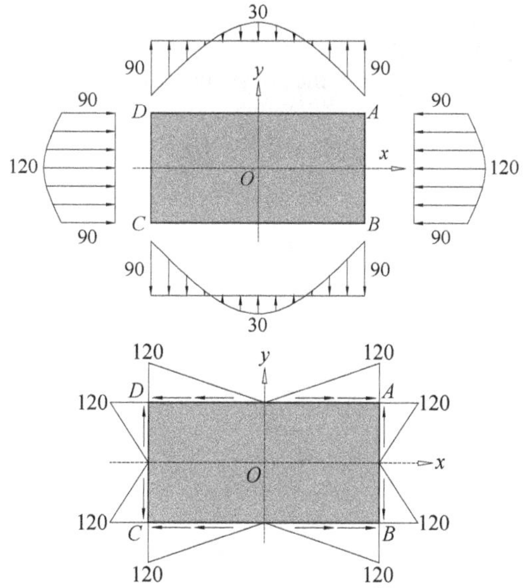

5.11. Determine the value of C in the Airy function of stresses indicated below that complies with the given boundary conditions on the top and bottom boundaries of the triangular plate shown in the following figure. Evaluate the components σ_x, τ_{xy} of the stress in the vertical section mn and trace the curves for $\alpha = 20°$.

$$\Phi = C\left[r^2(\alpha - \theta) + r^2 \sin\theta \cos\theta - r^2 \cos^2\theta \tan\alpha\right]$$

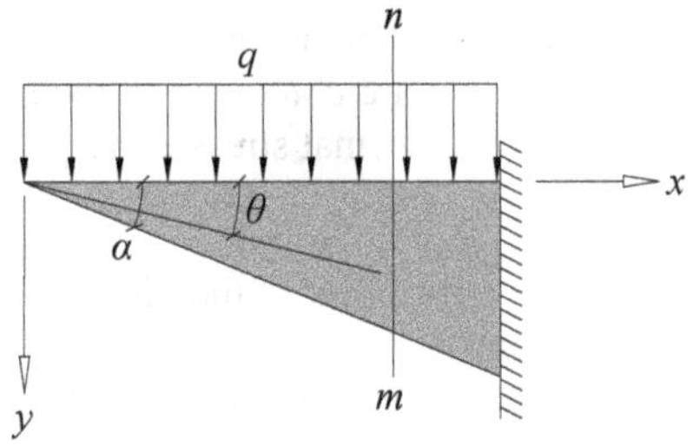

5.12. Find the stress field of a triangular dam with angle α in the vertex, under hydrostatic pressure from a liquid with density γ acting on the vertical face, and apply it to the case in which $\gamma = 20$ T/m^3 and $\alpha = 30°$. Verify that one can obtain it from the stress function:

$$\Phi = r^3\left[A\cos(3\theta) + B\sin(3\theta) + C\cos\theta + D\sin\theta\right]$$

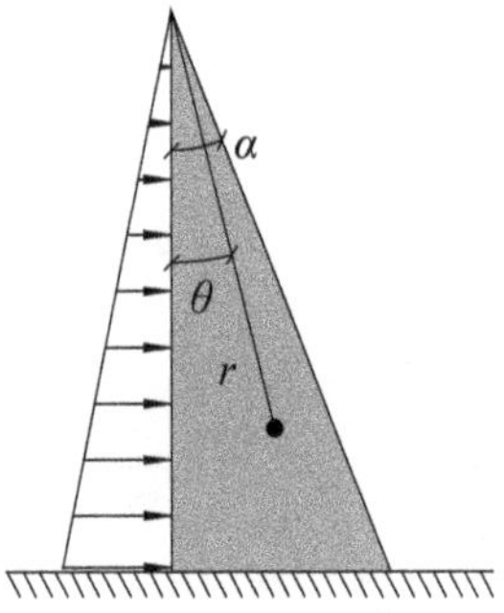

5.13. Demonstrate that the stress distribution given can be obtained from the stress function Φ that is indicated below and is the solution to the problem of a semi-infinite domain solicited by the vertical load p shown in the figure. The load extends indefinitely to the left.

$$\sigma_x = -\frac{p}{\pi}\left[\operatorname{atan}\left(\frac{y}{x}\right) + \frac{xy}{x^2+y^2}\right] \quad \sigma_y = -\frac{p}{\pi}\left[\operatorname{atan}\left(\frac{y}{x}\right) - \frac{xy}{x^2+y^2}\right]$$

$$\tau_{xy} = -\frac{p}{\pi}\frac{y^2}{x^2+y^2}$$

$$\Phi = -\frac{p}{2\pi}\left[\left(x^2+y^2\right)\operatorname{atan}\left(\frac{y}{x}\right) - xy\right] \quad \theta = \operatorname{atan}\left(\frac{y}{x}\right)$$

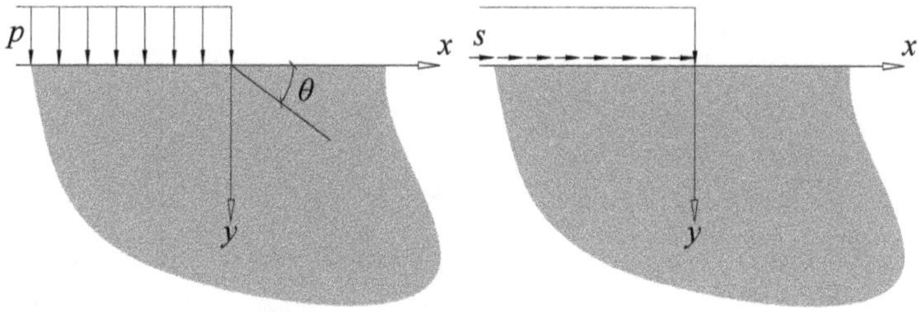

Also, study the value of τ_{xy} in the following cases:

1. At point P along the horizontal axis.
2. At point P along the vertical axis.

Similarly, demonstrate that the stress field of that same body solicited by the horizontal force s that appears in the figure can be obtained from the following stress functions:

$$\Phi = \frac{s}{\pi}\left[\frac{y^2 \ln\left(x^2+y^2\right)}{2} + xy\operatorname{atan}\left(\frac{y}{x}\right) - y^2\right]$$

Demonstrate that the stress σ_x becomes infinite at a point P that comes close to the coordinate's origin from any direction.

CHAPTER 6

HYPERELASTICITY

6.1 HYPERELASTIC MATERIALS

There are materials in which the relationship between stress and strain does not follow a linear law, that is to say, the strain is not proportional to the stress. The material, however, follows the same law in the loading and unloading process. Therefore, when the stress completely disappears, the strains also disappear. In general, this type of behaviour is common in materials with a high degree of deformability, such as the elastomers used to support bridge decks, in connection devices for aeronautical structures and in some mechanical components of the automobile industry. The behaviour of a linear and elastic material and that of a hyperelastic material are shown in figure 6.1.1.

One can recall that the expression for the differential increment dU of the strain energy is

$$dU = \sum_{i=1}^{3} \sigma_i d\varepsilon_i = \sum \frac{\partial U}{\partial \varepsilon_i} d\varepsilon_i \tag{6.1.1}$$

Therefore, the graphics of figure 6.1.1 can be represented on ε, σ axes or similarly on ε, $\partial U/\partial \varepsilon$ axes.

In a hyperelastic material, the expression for the strain energy U must comply with the following properties:

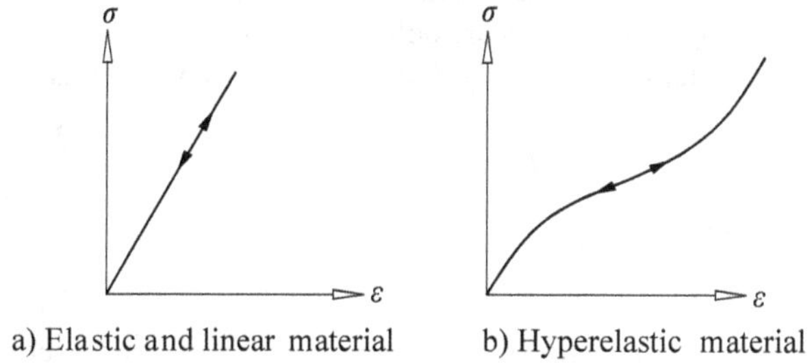

Figure 6.1.1: Materials with elastic and hyperelastic behaviour

1) After any number of loading and unloading cycles, if the strain is equal to zero, the strain energy will also be $U = 0$. No energy is stored, i.e.

$$\varepsilon = 0 \qquad U = 0 \tag{6.1.2a}$$

2) In the absence of strain, the stress is equal to zero, i.e.

$$\sigma\left(\varepsilon = 0\right) = \left(\frac{\partial U}{\partial \varepsilon}\right)_{\varepsilon=0} = 0 \tag{6.1.2b}$$

3) If the strain increases, so should the stress. Therefore, the following condition always applies

$$\left(\frac{\partial^2 U}{\partial \varepsilon^2}\right) > 0 \tag{6.1.2c}$$

In essence, the difference between a linearly elastic material and a hyperelastic material is that, in the first, the relationship between stress and strain is a constant value that corresponds to the modulus of elasticity E, and in the second, this relationship is variable and is set by the strain energy.

In hyperelastic materials, the strain energy is formulated as a function of the invariants of the Cauchy–Green tensors so as to be independent of the coordinate system that is being used. As will be shown, different expressions have been posed, and, in general, all of them are presented in the form

$$U = U\left(I_{C1}, I_{C2}, I_{C3}\right) \tag{6.1.3}$$

6.2 MODELS OF INCOMPRESSIBLE HYPERELASTIC MATERIALS

Elastomeric materials are used in engineering in a variety of applications, in bridge supports, connections between structural elements, machinery support or propulsion systems, amongst others. These types of materials suffer almost no volumetric changes when they are deformed; thus, they are often considered to be incompressible. Consequently, they will comply with equation (2.3.32); therefore

$$I_{C3} = \lambda_1^2 \lambda_2^2 \lambda_3^2 = 1 \tag{6.2.1}$$

and, according to (2.2.1.21)

$$J = 1 \tag{6.2.2}$$

For this type of material, various alternatives to the strain energy equation have been generated. Those that have proven to be more effective, and are therefore more commonly used, are described below.

6.2.1 Mooney–Rivlin model

This was one of the first models to be formulated to represent elastomeric materials. The strain energy is written as

$$U = \frac{\mu_1}{2}(I_{C1} - 3) - \frac{\mu_2}{2}(I_{C2} - 3) \qquad (6.2.1.1)$$

the reason for choosing this expression is because it is sufficient to consider two of the invariants because the third is related via (6.2.1). In addition, one can recall that

$$\mathbf{C}_r = 2\mathbf{E}_l + \mathbf{I} \qquad (6.2.1.2)$$

where $\mathbf{C}_r$ can be written as follows

$$\mathbf{C}_r = \begin{bmatrix} \lambda_1^2 & 0 & 0 \\ 0 & \lambda_2^2 & 0 \\ 0 & 0 & \lambda_3^2 \end{bmatrix} \qquad (6.2.1.3)$$

If there are no loads on the elastic body, and subsequently no strains, $\mathbf{C}_r = \mathbf{I}$ and therefore $\lambda_1 = \lambda_2 = \lambda_3 = 1$; then $I_{C1} = I_{C2} = 3$. The expression chosen for U in (6.2.1.1) ought to be equal to zero in the absence of loads, which is correct.

This model depends on two parameters, μ_1, μ_2, which are determined by the characteristics of each material. On occasions, equation (6.2.1.1) is modified and written as

$$U = c_1(I_{C1} - 3) + c_2(I_{C2} - 3) \qquad (6.2.1.4)$$

6.2.2 Neo-Hooke model

This is a simplification of the Mooney–Rivlin model, making $c_2 = 0$. In this case, the strain energy expression becomes

$$U = c(I_{C1} - 3) \qquad (6.2.2.1)$$

6.2.3 Yeoh model

In this model also, only the invariant I_{C1} is used; however, the expression is not linear, but cubic, therefore, the strain energy is expressed as

$$U = \sum_{i=1}^{3} c_i (I_{C1} - 3)^i \qquad (6.2.3.1)$$

On occasions, if the material's behaviour is different for compressive and tensile stresses, a different set of coefficients c_i $(i = 1, 2, 3)$ are used to express the strain energy for each class of stress.

6.2.4 Ogden model

In this model, the strain energy is formulated through polynomial expressions of the principal stresses, which leads to

$$U = \sum_{i=1}^{n} \frac{\mu_i}{\alpha_i} \left(\lambda_1^{\alpha_i} + \lambda_2^{\alpha_i} + \lambda_3^{\alpha_i} - 3\right) \tag{6.2.4.1}$$

where the values of μ_i, $\alpha_i\,(i = 1, \ldots, n)$ are experimentally obtained. In practice, this formulation is simplified by adopting only three terms of the summation. The most common values for elastomeric materials are

$$\alpha_1 = 1.3 \qquad \alpha_2 = 5 \qquad \alpha_3 = -2 \tag{6.2.4.2a}$$

$$\mu_1 = 630 \text{ kPa} \qquad \mu_2 = 1.2 \text{ kPa} \qquad \mu_3 = 10 \text{ kPa} \tag{6.2.4.2b}$$

One can observe that if the following values are adopted in equation (6.2.4.1) the model becomes that of Mooney–Rivlin.

$$n = 2 \qquad \alpha_1 = 2 \qquad \alpha_2 = -2 \tag{6.2.4.3}$$

6.3 STRAIN ENERGY IN HYPERELASTIC MATERIALS

One of the objectives in a hyperelastic material is to identify the stress tensor caused by the strains. Thus, and given that the materials are highly deformable, the strain energy equation (6.1.1) or the virtual work caused by some virtual strains can be expressed as follows.

For example, the virtual strain energy stored in a solid due to virtual strains $\delta\varepsilon_{eij}$ is expressed as

$$\delta U = \int t_{ij}\,\delta\varepsilon_{eij}\,dV_r \tag{6.3.1}$$

where t_{ij} are the components of the Cauchy stress tensor, $\delta\varepsilon_{eij}$ the components of the virtual Euler–Almansi strain tensor and dV_r the differential volume element after strain. Using the concept of double product or double contraction, this expression can be written as

$$\delta U = \int \boldsymbol{\tau} : \delta\mathbf{E}_e\,dV_r \tag{6.3.2}$$

where $\boldsymbol{\tau}$ is the Cauchy stress tensor and $\delta\mathbf{E}_e$ the Euler–Almansi virtual strain tensor. Recalling equation (2.2.1.21), one obtains

$$\delta U = \int \boldsymbol{\tau} : \delta\mathbf{E}_e J\,dV = \int J\boldsymbol{\tau} : \delta\mathbf{E}_e\,dV \tag{6.3.3}$$

and recalling (2.2.1.11)

$$\delta U = \int J\boldsymbol{\tau}:\mathbf{J}^{T^{-1}}\delta\mathbf{E}_l\mathbf{J}^{-1}dV \tag{6.3.4}$$

where $\delta\mathbf{E}_l$ is the Green–Lagrange virtual strain tensor. Using the attributes of the double product, the previous expression can be written as

$$\delta U = \int J\mathbf{J}^{-1}\boldsymbol{\tau}\mathbf{J}^{T^{-1}}:\delta\mathbf{E}_l dV \tag{6.3.5}$$

recalling equation (3.2.18) gives

$$\delta U = \int \boldsymbol{\tau}_{P2}:\delta\mathbf{E}_l dV \tag{6.3.6}$$

where $\boldsymbol{\tau}_{P2}$ is the second Piola–Kirchhoff tensor. Similarly to this expression, the differential strain energy dU per unit volume, that is to say, the variation of the energy for some differential strains defined by the tensor $d\mathbf{E}_l$, can be written as

$$dU = \boldsymbol{\tau}_{P2}:d\mathbf{E}_l \tag{6.3.7}$$

6.4 OBTAINING THE STRESS TENSOR IN HYPERELASTIC MATERIALS

The previous expression is very useful in the formulation of hyperelastic materials, i.e.

$$dU = \boldsymbol{\tau}_{P2}:d\mathbf{E}_l \tag{6.4.1}$$

reducing this into its components results in

$$dU = \sum t_{P2ij}de_{lij} \tag{6.4.2}$$

The differential strain energy can also be written as

$$dU = \sum \frac{\partial U}{\partial e_{lij}}de_{lij} \tag{6.4.3}$$

this means that

$$t_{P2ij} = \frac{\partial U}{\partial e_{lij}} \tag{6.4.4}$$

and, therefore

$$\boldsymbol{\tau}_{P2} = \frac{\partial U}{\partial \mathbf{E}_l} \tag{6.4.5}$$

recalling from (2.2.3.5) that

$$\mathbf{C}_r = 2\mathbf{E}_l + \mathbf{I} \tag{6.4.6}$$

where $\mathbf{C}_r$ is the direct Cauchy–Green strain tensor, results in

$$\tau_{P2} = \frac{\partial U}{\partial \mathbf{E}_l} = \frac{\partial U}{\partial \mathbf{C}_r}\frac{\partial \mathbf{C}_r}{\partial \mathbf{E}_l} = 2\frac{\partial U}{\partial \mathbf{C}_r} = 2\frac{\partial U}{\partial I_{C1}}\frac{\partial I_{C1}}{\partial \mathbf{C}_r} + 2\frac{\partial U}{\partial I_{C2}}\frac{\partial I_{C2}}{\partial \mathbf{C}_r} + 2\frac{\partial U}{\partial I_{C3}}\frac{\partial I_{C3}}{\partial \mathbf{C}_r} \tag{6.4.7}$$

in this expression, the derivatives with respect to $\mathbf{C}_r$ have been calculated via the invariants of this tensor. One can recall that

$$\mathbf{C}_r = \begin{bmatrix} \lambda_1^2 & 0 & 0 \\ 0 & \lambda_2^2 & 0 \\ 0 & 0 & \lambda_3^2 \end{bmatrix} \tag{6.4.8}$$

and

$$I_{C1} = \lambda_1^2 + \lambda_2^2 + \lambda_3^2 \qquad I_{C2} = \lambda_1^2\lambda_2^2 + \lambda_1^2\lambda_3^2 + \lambda_2^2\lambda_3^2 \qquad I_{C3} = \lambda_1^2\lambda_2^2\lambda_3^2 \tag{6.4.9}$$

from (6.4.8) and (6.4.9), one can verify that

$$\frac{\partial I_{C1}}{\partial \mathbf{C}_r} = \mathbf{I} \qquad \frac{\partial I_{C2}}{\partial \mathbf{C}_r} = I_{C1}\mathbf{I} - \mathbf{C}_r^T \qquad \frac{\partial I_3}{\partial \mathbf{C}_r} = I_{C3}\mathbf{C}_r^{-1^T} \tag{6.4.10}$$

By inserting these expressions into (6.4.7) and taking into account that the direct Cauchy–Green tensor is symmetric, the expression for the second Piola–Kirchhoff tensor becomes

$$\tau_{P2} = 2\left(\frac{\partial U}{\partial I_{C1}} + I_{C1}\frac{\partial U}{\partial I_{C2}}\right)\mathbf{I} - 2\frac{\partial U}{\partial I_{C2}}\mathbf{C}_r + 2I_{C3}\frac{\partial U}{\partial I_{C3}}\mathbf{C}_r^{-1} \tag{6.4.11}$$

recalling (3.2.20), one verifies that by premultiplying (6.4.11) by $\mathbf{J}$ and postmultiplying $\mathbf{J}^T$, one can obtain the Kirchhoff stress tensor

$$\tau_k = \mathbf{J}\tau_{P2}\mathbf{J}^T = 2\left(\frac{\partial U}{\partial I_{C1}} + I_{C1}\frac{\partial U}{\partial I_{C2}}\right)\mathbf{C}_l - 2\frac{\partial U}{\partial I_{C2}}\mathbf{C}_l^2 + 2I_{C3}\frac{\partial U}{\partial I_{C3}}\mathbf{I} \tag{6.4.12}$$

where $\mathbf{C}_l$ is the Cauchy–Green left tensor. From this, it is easy to obtain the Cauchy stress tensor τ, which was the desired objective, i.e.

$$\tau = \frac{1}{J}\tau_k = \frac{2}{J}\left[\left(\frac{\partial U}{\partial I_{C1}} + I_{C1}\frac{\partial U}{\partial I_{C2}}\right)\mathbf{C}_l - \frac{\partial U}{\partial I_{C2}}\mathbf{C}_l^2 + I_{C3}\frac{\partial U}{\partial I_{C3}}\mathbf{I}\right] \tag{6.4.13}$$

6.5 STRESS TENSORS IN INCOMPRESSIBLE HYPERELASTIC MATERIALS. UNIAXIAL FORCES

It is interesting to obtain the expression for the stresses in these materials when they are solicited by specific loads. One of them is the case in which the forces exist exclusively in one direction.

If one applies a load F to a specimen with area A and length l, the new values will be A_r and l_r after the strain. From the volume conservation hypothesis, the stretch value λ can be determined as

$$\lambda = \frac{l_r}{l} = \frac{A}{A_r} \tag{6.5.1}$$

calling this stretch λ_1, the stretches in the other directions are

$$\lambda_2 = \lambda_3 = \frac{1}{\sqrt{\lambda}} \tag{6.5.2}$$

The invariants will be

$$I_{C1} = \lambda^2 + \frac{2}{\lambda} \qquad I_{C2} = 2\lambda + \frac{1}{\lambda^2} \qquad I_{C3} = 1 \tag{6.5.3}$$

As forces only exist in one direction, the associated stresses will be zero. That is to say, one has

$$\sigma_1 = \sigma \qquad \sigma_2 = \sigma_3 = 0 \tag{6.5.4}$$

The expression for the Cauchy stress tensor becomes

$$\boldsymbol{\tau} = 2\left(\frac{\partial U}{\partial I_{C1}} + I_{C1}\frac{\partial U}{\partial I_{C2}}\right)\mathbf{C}_l - 2\frac{\partial U}{\partial I_{C2}}\mathbf{C}_l^2 + 2I_{C3}\frac{\partial U}{\partial I_{C3}}\mathbf{I} \tag{6.5.5}$$

and therefore the stress tensor is

$$\begin{bmatrix} \sigma & 0 & 0 \\ 0 & 0 & 0 \\ 0 & 0 & 0 \end{bmatrix} = 2\left[\frac{\partial U}{\partial I_{C1}} + \left(\lambda^2 + \frac{2}{\lambda}\right)\frac{\partial U}{\partial I_{C2}}\right]\begin{bmatrix} \lambda^2 & 0 & 0 \\ 0 & \frac{1}{\lambda} & 0 \\ 0 & 0 & \frac{1}{\lambda} \end{bmatrix}$$

$$-\frac{2\partial U}{\partial I_{C2}}\begin{bmatrix} \lambda^4 & 0 & 0 \\ 0 & \frac{1}{\lambda^2} & 0 \\ 0 & 0 & \frac{1}{\lambda^2} \end{bmatrix} + \frac{2\partial U}{\partial I_{C3}}\mathbf{I} \tag{6.5.6}$$

resulting in the following equations

$$\sigma = 2\lambda^2\frac{\partial U}{\partial I_{C1}} + 2\lambda^2\left(\lambda^2 + \frac{2}{\lambda}\right)\frac{\partial U}{\partial I_{C2}} - 2\lambda^4\frac{\partial U}{\partial I_{C2}} + 2\frac{\partial U}{\partial I_{C3}} \tag{6.5.7a}$$

$$0 = \frac{2}{\lambda}\frac{\partial U}{\partial I_{C1}} + \frac{2}{\lambda}\left(\lambda^2 + \frac{2}{\lambda}\right)\frac{\partial U}{\partial I_{C2}} - \frac{2}{\lambda^2}\frac{\partial U}{\partial I_{C2}} + 2\frac{\partial U}{\partial I_{C3}} \tag{6.5.7b}$$

$$0 = \frac{2}{\lambda}\frac{\partial U}{\partial I_{C1}} + \frac{2}{\lambda}\left(\lambda^2 + \frac{2}{\lambda}\right)\frac{\partial U}{\partial I_{C2}} - \frac{2}{\lambda^2}\frac{\partial U}{\partial I_{C2}} + 2\frac{\partial U}{\partial I_{C3}} \tag{6.5.7c}$$

Notice that equations (6.5.7b) and (6.5.7c) are identical.

From equation (6.5.7b), one can obtain the expression for $\partial U/\partial I_{C3}$ in the function of the other derivatives of the invariants, and replacing this in (6.5.7a), one finally obtains

$$\sigma = 2\left(\lambda^2 - \frac{1}{\lambda}\right)\left(\frac{\partial U}{\partial I_{C1}} + \frac{1}{\lambda}\frac{\partial U}{\partial I_{C2}}\right) \tag{6.5.8}$$

The particularization of this stress expression can then be obtained for each of the models of hyperelastic materials previously presented.

6.5.1 Stress in the Mooney–Rivlin model

Substituting in equation (6.5.8) the result corresponding to the strain energy expression defined in (6.2.1.4), the stress is written as

$$\sigma = 2\left(\lambda^2 - \frac{1}{\lambda}\right)\left(c_1 + \frac{c_2}{\lambda}\right) \tag{6.5.1.1}$$

6.5.2 Stress in the Neo-Hooke model

Similarly, in this case, the expression for the stress is

$$\sigma = 2\left(\lambda^2 - \frac{1}{\lambda}\right)c \tag{6.5.2.1}$$

6.5.3 Stress in the Yeoh model

Assuming that the same set of coefficients are used for compressive and tensile stresses, the resulting expression for the stress is

$$\sigma = 2\left(\lambda^2 - \frac{1}{\lambda}\right)\left[c_1 + 2c_2\left(I_{C1} - 3\right) + 3c_3\left(I_{C1} - 3\right)^2\right] \tag{6.5.3.1}$$

with

$$I_{C1} = \lambda^2 + \frac{2}{\lambda} \tag{6.5.3.2}$$

6.5.4 Stress in the Ogden model

In this case, as the expression that defines the strain energy is a function of the stretch, one should perform an intermediate step. Starting with (6.5.8), i.e.

$$\sigma = 2\left(\lambda^2 - \frac{1}{\lambda}\right)\left(\frac{\partial U}{\partial \lambda}\frac{\partial \lambda}{\partial I_{C1}} + \frac{1}{\lambda}\frac{\partial U}{\partial \lambda}\frac{\partial \lambda}{\partial I_{C2}}\right) \tag{6.5.4.1}$$

and recalling that

$$I_{C1} = \lambda^2 + \frac{2}{\lambda} \quad I_{C2} = 2\lambda + \frac{1}{\lambda^2} \tag{6.5.4.2}$$

results in

$$\frac{\partial \lambda}{\partial I_{C1}} = \frac{\lambda^2}{2\lambda^3 - 2} \quad \frac{\partial \lambda}{\partial I_{C2}} = \frac{\lambda^3}{2\lambda^3 - 2} \tag{6.5.4.3}$$

referring back to (6.5.4.1), one finally has the following formula for the Cauchy stress

$$\sigma = 2\lambda \frac{\partial U}{\partial \lambda} \tag{6.5.4.4}$$

recalling now the expressions for U given by (6.2.4.1) and for the stretches given by (6.5.2) gives

$$U = \sum_{i=1}^{n} \frac{\mu_i}{\alpha_i} \left(\lambda^{\alpha_i} + 2\lambda^{-0.5\alpha_i} - 3 \right) \tag{6.5.4.5}$$

substituting in (6.5.4.4), one finally obtains

$$\sigma = 2 \sum_{i=1}^{n} \mu_i \left(\lambda^{\alpha_i} - \lambda^{-0.5\alpha_i} \right) \tag{6.5.4.6}$$

6.6 COMPRESSIBLE HYPERELASTIC MATERIALS

Sometimes the incompressibility hypothesis of materials is not appropriate. This happens in the case of some specific types of elastomers and biological materials. The study of the latter type of materials has been the subject of growing interest due to the development of bioengineering, which, among other things, focuses on the mechanical behaviour of the tissues of living organisms. When the material is compressible, one should eliminate the condition that the volume remains constant, and consequently the equations (6.2.1) and (6.2.2) are not satisfied, that is to say

$$I_{C3} \neq 1 \quad J \neq 1 \tag{6.6.1}$$

equations (6.4.11), (6.4.12) and (6.4.13) for the stress tensors are still valid for these materials, however.

The following paragraphs describe some of the models used for compressible hyperelastic materials and the expressions for the strain energy adopted. From these, the different stress tensors can be obtained from equations (6.4.11) to (6.4.13) already mentioned.

6.6.1 Mooney–Rivlin model

In this model of compressible hyperelastic materials, a new term is added to the strain energy expressions presented in (6.2.1.1) and (6.2.1.4), which leads to

$$U = c_1 \left(I_{C1} - 3\right) + c_2 \left(I_{C2} - 3\right) + d \left(J - 1\right)^2 \tag{6.6.1.1}$$

as the material is compressible, the last term on the right-hand side cannot be eliminated. However, when the loads are not present and, consequently, strains are zero, all of the terms become zero; thus, the strain energy does not exist.

To operate with this expression, it is necessary to relate the determinant J to the invariants of the Cauchy–Green strain tensor. One can recall that

$$\tau_{P2} = 2\left(\frac{\partial U}{\partial I_{C1}} + I_{C1}\frac{\partial U}{\partial I_{C2}}\right)\mathbf{I} - 2\frac{\partial U}{\partial I_{C2}}\mathbf{C}_r + 2I_{C3}\frac{\partial U}{\partial I_{C3}}\mathbf{C}_r^{-1} \tag{6.6.1.2}$$

and that

$$J = |\mathbf{J}| \tag{6.6.1.3}$$

it is also known that

$$\mathbf{C}_r = \mathbf{J}^T \mathbf{J} \tag{6.6.1.4}$$

and that the determinants of the matrices satisfy the following characteristic

$$|\mathbf{C}_r| = |\mathbf{J}^T|\,|\mathbf{J}| = J^2 \tag{6.6.1.5}$$

therefore

$$\lambda_1^2\lambda_2^2\lambda_3^2 = J^2 = I_{C3} \tag{6.6.1.6}$$

As a result, the strain energy can be expressed as

$$U = c_1 \left(I_{C1} - 3\right) + c_2 \left(I_{C2} - 3\right) + d\left(\sqrt{I_{C3}} - 1\right)^2 \tag{6.6.1.7}$$

from this, one applies equation (6.4.13) to obtain the Cauchy strain tensor

$$\tau = \frac{2}{J}\left[\left(\frac{\partial U}{\partial I_{C1}} + I_{C1}\frac{\partial U}{\partial I_{C2}}\right)\mathbf{C}_l - \frac{\partial U}{\partial I_{C2}}\mathbf{C}_l^2 + I_{C3}\frac{\partial U}{\partial I_{C3}}\mathbf{I}\right] \tag{6.6.1.8}$$

substituting the corresponding derivatives gives

$$\tau = \frac{2}{J}\left[\left(c_1 + I_{C1}c_2\right)\mathbf{C}_l - c_2\mathbf{C}_l^2 + Jd\left(J - 1\right)\mathbf{I}\right] \tag{6.6.1.9}$$

6.6.2 Neo-Hooke model

As previously mentioned, this model results from the simplification of the previous one by making $c_2 = 0$. Consequently, the strain energy expression becomes

$$U = c\,(I_{C1} - 3) + d\left(\sqrt{I_{C3}} - 1\right)^2 \qquad (6.6.2.1)$$

and the Cauchy strain tensor results from (6.6.1.9), i.e.

$$\boldsymbol{\tau} = \frac{2}{J}\,[c\mathbf{C}_l + Jd\,(J - 1)\,\mathbf{I}] \qquad (6.6.2.2)$$

Another more general formulation is defined for some materials using the following expression for the strain energy

$$U = c\,(I_{C1} - 3) - 2c\ln J + d\,(\ln J)^2 \qquad (6.6.2.3)$$

substituting the corresponding derivatives from (6.6.1.8), one finally obtains

$$\boldsymbol{\tau} = \frac{2}{J}\,[c\,(\mathbf{C}_l - \mathbf{I}) + d\ln J\mathbf{I}] \qquad (6.6.2.4)$$

6.6.3 Ogden model

When the condition of incompressibility is eliminated in this model, equation (6.2.4.1) includes a new term and becomes

$$U = \sum_{i=1}^{n} \frac{\mu_i}{\alpha_i}\left(\lambda_1^{\alpha_i} + \lambda_2^{\alpha_i} + \lambda_3^{\alpha_i} - 3\right) + d\,(J - 1)^2 \qquad (6.6.3.1)$$

Remember that if this expression is particularized for

$$n = 2 \quad \alpha_1 = 2 \quad \alpha_2 = -2 \qquad (6.6.3.2)$$

the Mooney–Rivlin model is obtained.

6.6.4 Yeoh model

For compressible hyperelastic materials, equation (6.2.3.1) expressing the strain energy is transformed into

$$U = \sum_{i=1}^{3} c_i\,(I_{C1} - 3)^i + \sum_{i=1}^{3} d_i\,(J - 1)^{2i} \qquad (6.6.4.1)$$

using equation (6.6.1.6) results in

$$U = \sum_{i=1}^{3} c_i (I_{C1} - 3)^i + \sum_{i=1}^{3} d_i \left(\sqrt{I_{C3}} - 1\right)^{2i} \tag{6.6.4.2}$$

From (6.4.13), the formulation for the Cauchy stress tensor becomes

$$\begin{aligned} \boldsymbol{\tau} = \frac{2}{J}\Big\{ & \left[c_1 + 2c_2 (I_{C1} - 3) + 3c_3 (I_{C1} - 3)^2\right] \mathbf{C}_l \\ & + J\left[d_1 (J - 1) + 2d_2 (J - 1)^3 + 3d_3 (J - 1)^5\right] \mathbf{I} \Big\} \end{aligned} \tag{6.6.4.3}$$

EXERCISES

6.1. Consider a thin sheet of an incompressible hyperelastic material inserted into a rigid rectangular frame. Two sets of loads are applied to produce the stretches that appear in the following figures. Obtain the expression for the Cauchy stress tensor according to the strain energy in each of these cases:

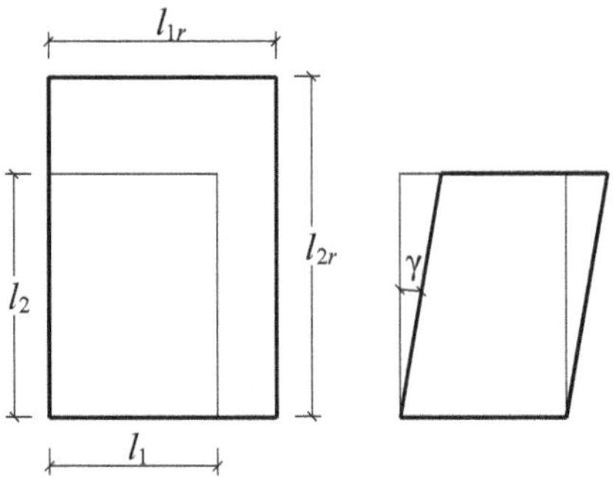

6.2. Obtain the final expression for the Cauchy stress tensor for the loads from the previous exercise, assuming that:

1. The material is behaving according to the Mooney–Rivlin model.
2. The material is behaving according to the Yeoh model.

6.3. In a test of uniaxial stress of an incompressible hyperelastic material, one has obtained the curve in the figure that relates the stress to the stretch of the specimen. Obtain the analytical expression for the Cauchy stress for each of the following models:

1) Neo-Hooke model.
2) Mooney–Rivlin model.

3) Yeoh model.
4) Ogden model.

λ	σ [MPa]
1.036	0.065
1.071	0.110
1.107	0.165
1.143	0.222
1.178	0.276
1.222	0.352
1.251	0.411
1.285	0.481
1.329	0.578
1.357	0.635
1.392	0.722
1.429	0.831
1.464	0.931
1.508	1.065
1.536	1.166
1.571	1.281
1.607	1.405
1.643	1.551
1.686	1.722
1.721	1.874
1.752	2.015
1.793	2.220
1.829	2.411
1.858	2.554
1.893	2.757
1.928	2.971
1.964	3.194
2.007	3.456
2.043	3.691
2.072	3.892
2.106	4.144
2.143	4.401
2.179	4.654

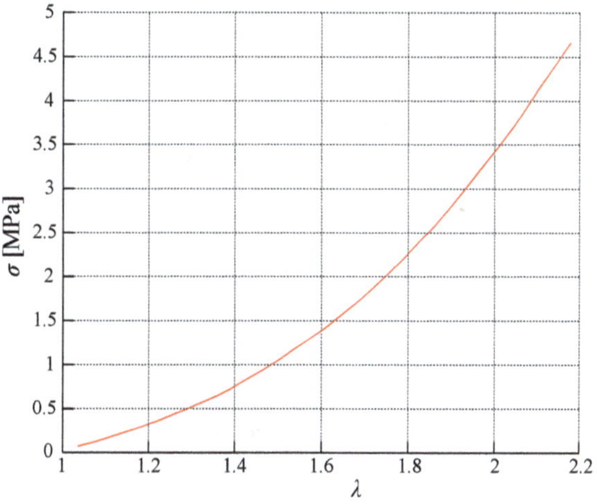

CHAPTER 7

PLASTICITY

7.1 INTRODUCTION

Some materials, after a linear or non-linear elastic phase, in which the strain is recovered when the acting force disappears, show a different behaviour characterized by a sudden and continuous increment in the strains, without a stress increase being required. This behaviour is referred to as *plastic*, and the strains produced in this phase do not recover when the material is unloaded; they become permanent. In a new loading cycle, the elastic phase can exist in the same stress range or it can increase; that is to say, an elastic behaviour with a higher range of stress values than that of the initial loading cycle can exist. This behaviour is known as *strain hardening*. Figures 7.1.1a and 7.1.1b describe each of these cases.

7.2 PLASTIFICATION CRITERIA IN METALLIC MATERIALS

7.2.1 Beltrami–Haigh criterion

This criterion is based on the concept that the elastic phase terminates when the strain energy stored in the material reaches a certain value. As a result,

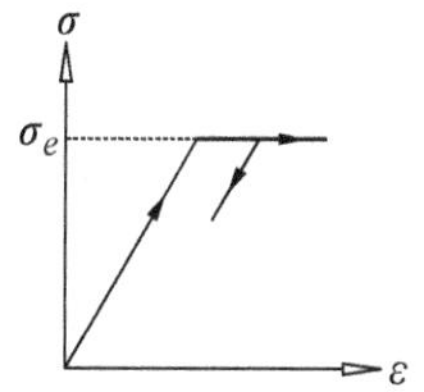

a) Elastoplastic behaviour without strain hardening

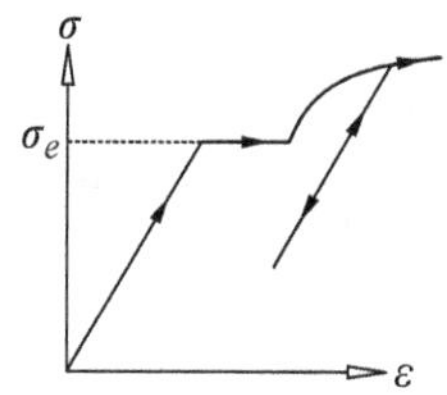

b) Elastoplastic behaviour with strain hardening

Figure 7.1.1: Materials with elastoplastic behaviour

plasticity is achieved in the principal stress and strain axes when

$$U = \frac{1}{2}\sum_{i=1}^{3}\sigma_i\varepsilon_i = k^2 \tag{7.2.1.1}$$

i.e. the strain energy reaches a value k^2, which is unique to each material.

If the strain energy is expressed only as a function of the stresses, one obtains

$$U = \frac{1}{2E}\left[\sigma_1^2 + \sigma_2^2 + \sigma_3^2 - 2\nu\left(\sigma_1\sigma_2 + \sigma_1\sigma_3 + \sigma_2\sigma_3\right)\right] = k^2 \tag{7.2.1.2}$$

In a tensile test, the elastic phase ends for $\sigma_1 = \sigma_e$, $\sigma_2 = \sigma_3 = 0$; therefore, the value of k^2 can be written as

$$k^2 = \frac{\sigma_e^2}{2E} \tag{7.2.1.3}$$

In simple shear stress, one can write $\tau = \sigma_1 = -\sigma_3$, $\sigma_2 = 0$. Entering these values in (7.2.1.2) gives

$$\frac{2\tau^2 + 2\nu\tau^2}{2E} = \frac{\tau^2(1+\nu)}{E} = \frac{\sigma_e^2}{2E} \tag{7.2.1.4}$$

or

$$\tau = \frac{\sigma_e}{\sqrt{2(1+\nu)}} \tag{7.2.1.5}$$

For the case of steel, $\nu = 0.3$; thus, the above formula gives $\tau = 0.62\sigma_e$. This value, however, does not coincide with experimental results, which show that $\tau \approx 0.56\sigma_e$.

The expression for the surface of plastification, which is the locus of the points that represent the stress vector at the end of the elastic phase, can be obtained by substituting (7.2.1.3) in (7.2.1.2), as shown below

$$\sigma_1^2 + \sigma_2^2 + \sigma_3^2 - 2\nu\left(\sigma_1\sigma_2 + \sigma_1\sigma_3 + \sigma_2\sigma_3\right) = \sigma_e^2 \tag{7.2.1.6}$$

which corresponds to an ellipsoid whose longest axis is the principal diagonal of the first quadrant. In the σ_1, σ_2 axes, this produces the ellipse of figure 7.2.1.1b and, from the Haigh–Westergaard plane, one obtains the circle with radius $\sigma_e/\sqrt{1+\nu}$.

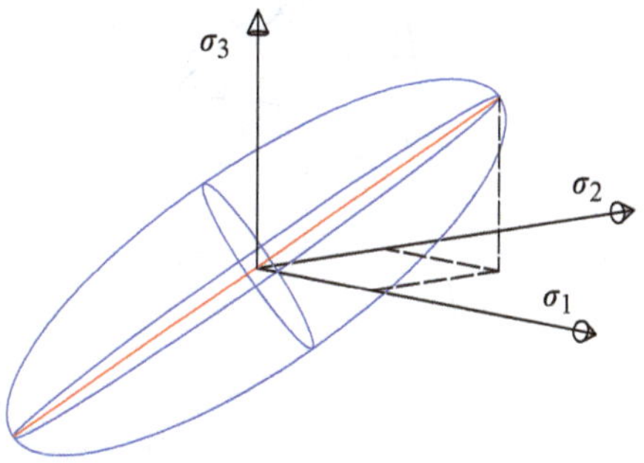

a) Ellipsoid

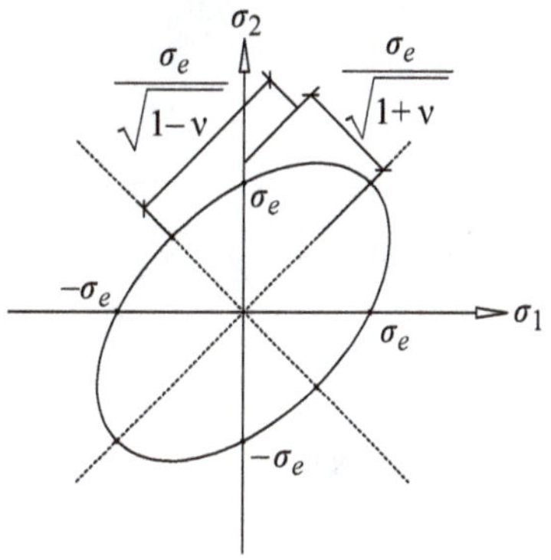

b) Section in σ_1, σ_2 plane

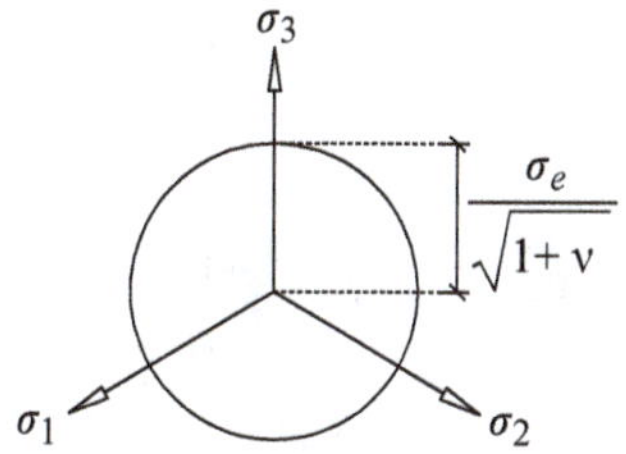

c) Section of the Haigh–Westergaard plane

Figure 7.2.1.1: Representation of the Beltrami–Haigh criterion

7.2.2 Von Mises–Hencky criterion

This criterion modifies the previous one, proposing that the most important aspect of material plastification is the strain energy due to the deviatoric tensor and not the overall strain energy. This means that one should subtract that corresponding to the spherical tensor from the strain energy expression defined by equation (7.2.1.1), thus obtaining

$$U_m = \frac{3}{2}\sigma_m \varepsilon_m = \frac{3}{2}\left(\frac{\sigma_1 + \sigma_2 + \sigma_3}{3}\right)\left(\frac{\varepsilon_1 + \varepsilon_2 + \varepsilon_3}{3}\right) \tag{7.2.2.1}$$

the expression for ε_m in the function of the principal stresses can be written as

$$\varepsilon_m = \frac{1 - 2\nu}{3E}(\sigma_1 + \sigma_2 + \sigma_3) \tag{7.2.2.2}$$

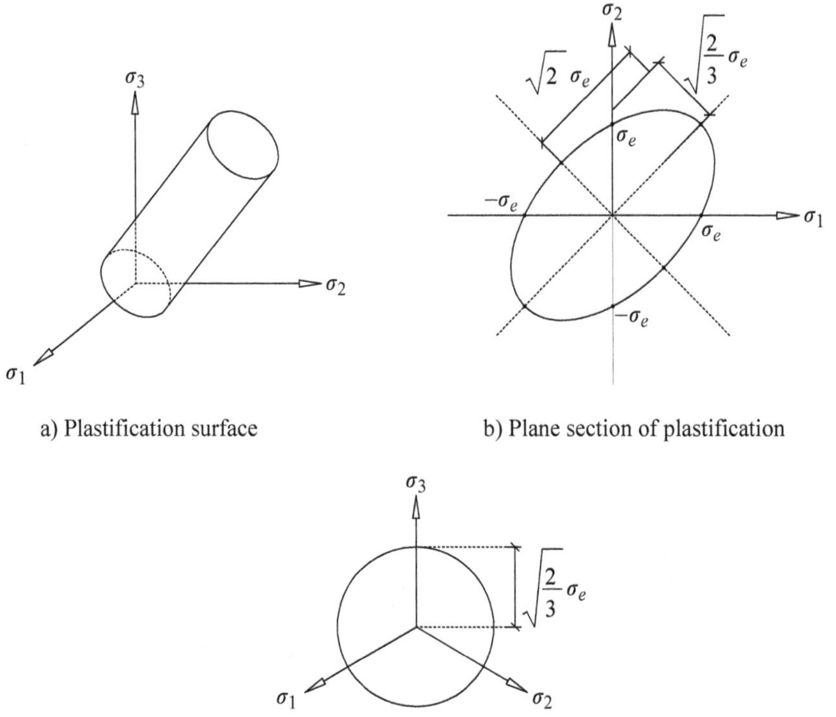

a) Plastification surface

b) Plane section of plastification

c) Haigh–Westergaard representation

Figure 7.2.2.1: Representation of the Von Mises–Hencky criterion

resulting in

$$U_m = \frac{1-2\nu}{6E}\,(\sigma_1 + \sigma_2 + \sigma_3)^2 \qquad (7.2.2.3)$$

consequently, and recalling (7.2.1.2), one obtains

$$U_d = U - U_m = \frac{1+\nu}{6E}\left[(\sigma_1 - \sigma_2)^2 + (\sigma_1 - \sigma_3)^2 + (\sigma_2 - \sigma_3)^2\right] \qquad (7.2.2.4)$$

The criterion of Von Mises–Hencky establishes that plastification occurs when U_d reaches a limit value, different for each material. That value can be easily obtained from a tensile test in which this phenomenon appears at the tensile stress of plastification σ_e, i.e. $\sigma_1 = \sigma_e$ and $\sigma_2 = \sigma_3 = 0$. Entering these values into equation (7.2.2.4) gives the maximum value of U_d as

$$U_d = \frac{1+\nu}{3E}\sigma_e^2 \tag{7.2.2.5}$$

substituting this into (7.2.2.4) gives

$$\sigma_e = \frac{1}{\sqrt{2}}\sqrt{(\sigma_1-\sigma_2)^2+(\sigma_1-\sigma_3)^2+(\sigma_2-\sigma_3)^2} \tag{7.2.2.6}$$

this is the expression for the plastification criterion.

The relationship between the Von Mises criterion and the deviatoric tensor can be obtained by developing the right-hand side of equation (7.2.2.6) and squaring it. One can then write

$$(\sigma_1+\sigma_2+\sigma_3)^2 - 3(\sigma_1\sigma_2+\sigma_1\sigma_3+\sigma_2\sigma_3) = J_1^2 - 3J_2 \tag{7.2.2.7}$$

recalling the expression for the invariants of the deviatoric tensor, i.e.

$$J_1^2 - 3J_2 = -3J_{2d} \tag{7.2.2.8}$$

the Von Mises–Hencky criterion can be expressed as

$$\sqrt{-3J_{2d}} = \sigma_e \tag{7.2.2.9}$$

indicating that plastification starts when the value of the second invariant of the deviatoric tensor reaches a certain value.

This criterion is also associated with other factors. If one calculates the shear component of the octahedral stress, that is to say, the stress vector that corresponds to a plane that forms equal angles with the principal directions, one obtains

$$\tau_{oct} = \frac{1}{3}\sqrt{(\sigma_1-\sigma_2)^2+(\sigma_1-\sigma_3)^2+(\sigma_2-\sigma_3)^2} \tag{7.2.2.10}$$

which implies that

$$\tau_{oct} = \frac{\sqrt{2}}{3}\sigma_e \tag{7.2.2.11}$$

this implies that plastification appears when the shear component of the octahedral stress has a certain value. This fact also explains why other plastification criteria exist, based on studies of the maximum value of the shear stress, as will be seen later.

A shear stress field τ is equivalent to another in which $\tau = \sigma_1 = -\sigma_3$, $\sigma_2 = 0$. Substituting these values in (7.2.2.6), one finally obtains

$$\tau = \frac{\sigma_e}{\sqrt{3}} = \sqrt{-J_{2d}} \tag{7.2.2.12}$$

therefore, the maximum value of shear stress is $\tau = 0.577\sigma_e$, which is similar to the experimental result $\tau \approx 0.56\sigma_e$.

The representation of the Von Mises–Hencky criterion, defined by equation (7.2.2.6), is an infinite cylinder whose directrix forms equal angles with the first quadrant. A plane section produces an ellipse that cuts the coordinate axes at a distance σ_e from the coordinates' origin, and their representation in the Haigh–Westergaard plane is a circle with radius $\sqrt{2/3}\sigma_e$, as shown in figure 7.2.2.1.

7.2.3 Tresca criterion

This criterion is not based on energy approaches, but establishes that plastification occurs when shear stress reaches the maximum value admissible by the material. In any two-dimensional stress field, one can verify that the maximum shear stress, as shown in figure 7.2.3.1, is

$$\tau_{\max} = \left| \frac{\sigma_i - \sigma_j}{2} \right| \tag{7.2.3.1}$$

If the field is uniaxial tensile, the maximum value of the shear stress is

$$\tau_{\max} = \frac{\sigma_1}{2} \tag{7.2.3.2}$$

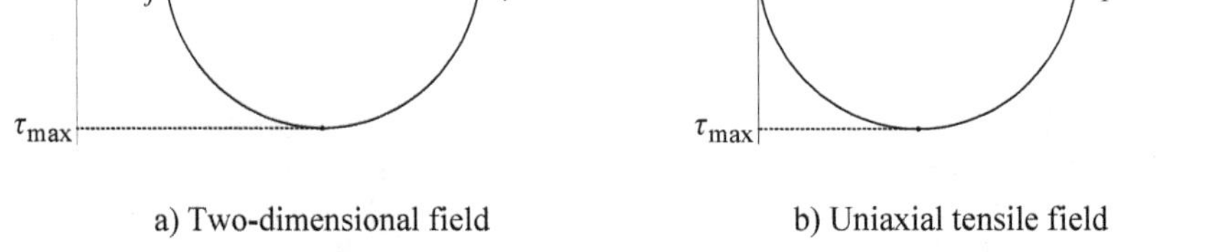

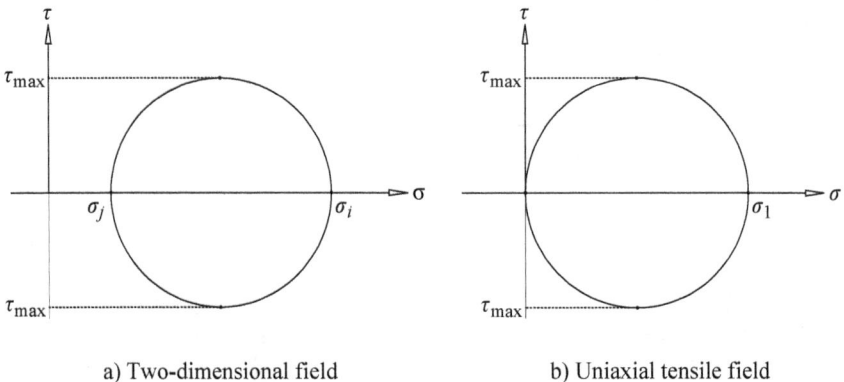

Figure 7.2.3.1: Maximum shear stresses

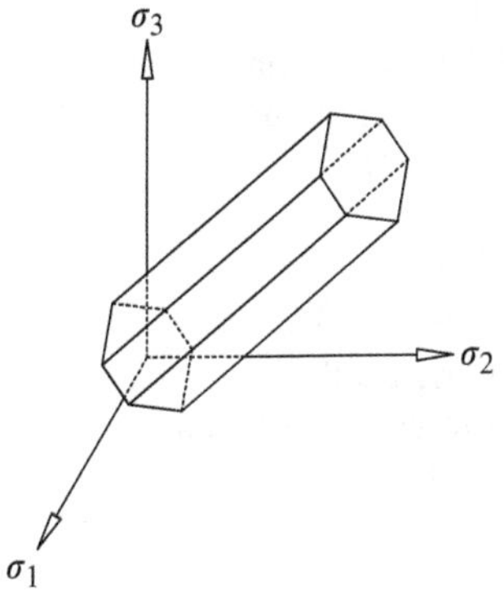

a) Plastification surface

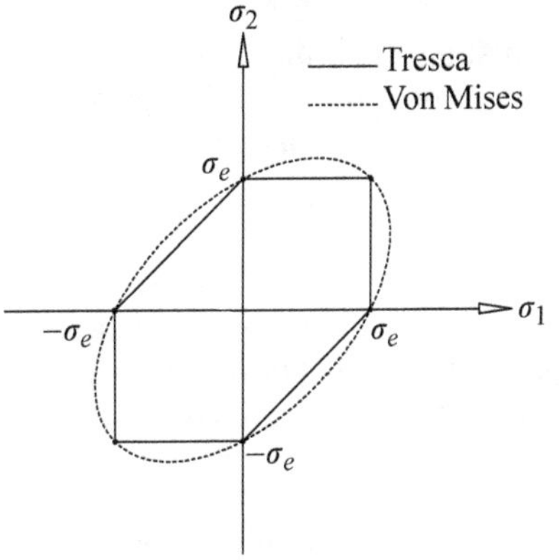

b) Section in σ_1, σ_2 plane

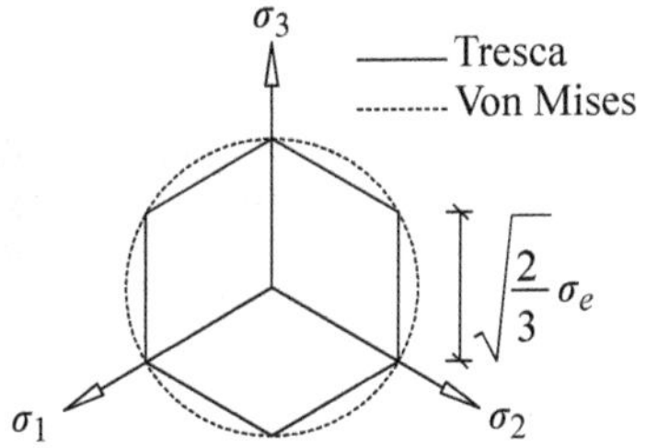

c) Section of the Haigh–Westergaard plane

Figure 7.2.3.2: Representation of the Tresca criterion

If σ_e is the stress at which the material plastifies, the maximum shear stress that characterises the material will therefore be

$$\tau_{\max} = \frac{\sigma_e}{2} \tag{7.2.3.3}$$

Therefore, the plasticity criterion will consist of six expressions resulting from (7.2.3.1), which are

$$\sigma_1 - \sigma_2 = \pm\sigma_e \quad \sigma_1 - \sigma_3 = \pm\sigma_e \quad \sigma_2 - \sigma_3 = \pm\sigma_e \tag{7.2.3.4}$$

They constitute a hexagonal prism whose directrix is the straight line of the first quadrant that forms equal angles with the three coordinate axes. In a plane section, the prism brings forth the hexagon from figure 7.2.3.2b, and in the Haigh–Westergaard plane it has the shape of a regular hexagon with sides measuring $\sqrt{2/3}\sigma_e$.

7.3 PLASTIFICATION CRITERIA IN SOIL MECHANICS

This section will describe several plastification criteria that can be applied to soil mechanics. In this field of engineering, compressive stresses are considered positive since they are the ones that these materials are subjected to. The positive axis of the normal stresses will correspond to compressive ones.

7.3.1 Mohr–Coulomb criterion

This criterion is applicable mainly to rocks and granular materials. It is based on the principle that the shear stress is what is important. Thus, it establishes that for a two-dimensional stress field defined by Mohr's circle, as shown in figure 7.3.1.1, the plasticity condition will be as follows

$$\tau = c + \sigma \tan \phi \tag{7.3.1.1}$$

where τ, σ are the stress components on a point and c, ϕ are the cohesion and the angle of internal friction, respectively, which are characteristic values of each material. One can verify that this criterion is a generalization of the one by Tresca when

$$c = \frac{\sigma_e}{2} \quad \phi = 0 \tag{7.3.1.2}$$

At point A of the figure, the values of the stress components are

$$\tau_A = r\cos\phi \quad \sigma_A = \frac{\sigma_i + \sigma_j}{2} - r\sin\phi \tag{7.3.1.3}$$

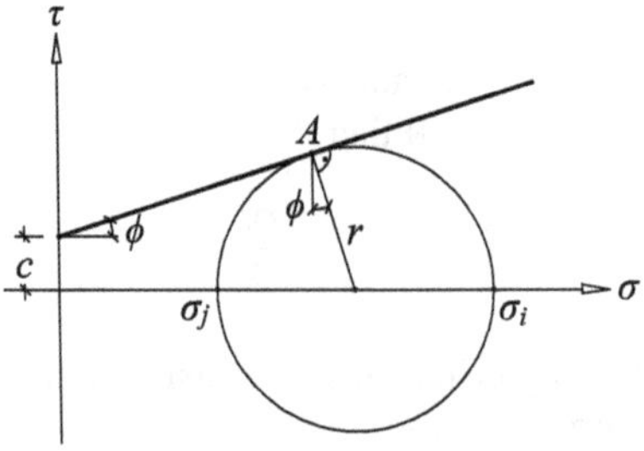

Figure 7.3.1.1: Mohr–Coulomb plasticity criterion

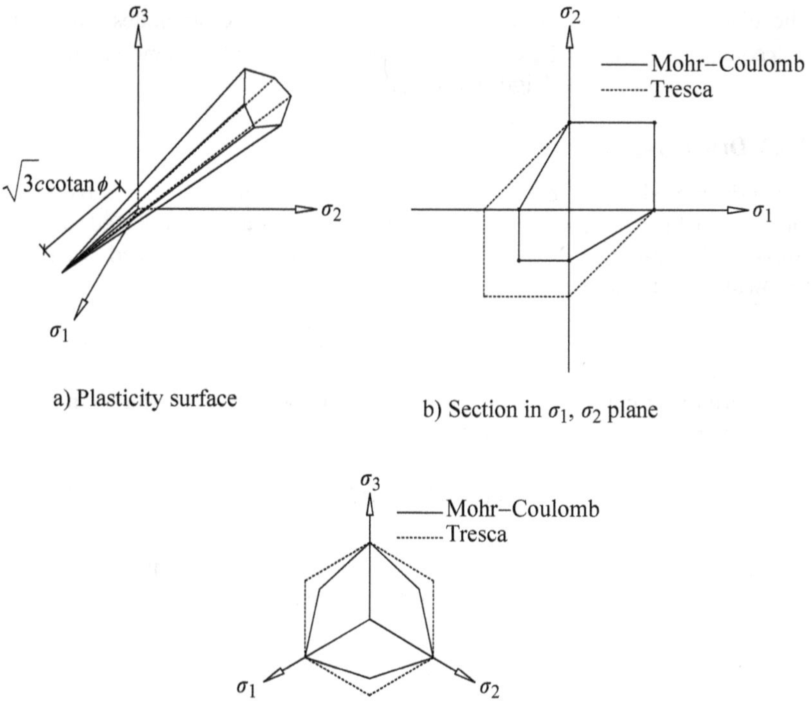

a) Plasticity surface

b) Section in σ_1, σ_2 plane

c) Haigh–Westergaard representation

Figure 7.3.1.2: Mohr–Coulomb criterion

substituting in (7.3.1.1) and noting that

$$r = \frac{\sigma_i - \sigma_j}{2} \tag{7.3.1.4}$$

results in

$$\sigma_i - \sigma_j - \left(\sigma_i + \sigma_j\right) \sin\phi - 2c\cos\phi = 0 \tag{7.3.1.5}$$

which is the condition of plasticity. This equation can be employed for any combination of two-dimensional stress fields, resulting in the following equation

$$\left[\pm\sigma_i - \left(\pm\sigma_j\right)\right] - \left[\pm\sigma_i + \left(\pm\sigma_j\right)\right] \sin\phi - 2c\cos\phi = 0 \tag{7.3.1.6}$$

This equation produces a surface that is a hexagonal pyramid whose axis is the straight line which forms equal angles with the principal directions.

The distance from the vertex of the pyramid to the coordinates' origin is $\sqrt{3}c\text{cotan}\phi$. Figure 7.3.1.2 shows the plasticity surface, a plane section and its representation in the Haigh–Westergaard space.

7.3.2 Drucker–Prager criterion

This criterion also applies to rocks and granular materials, but it is based on the relationship between the octahedral normal stress and the octahedral shear stress, and it establishes that the material's plasticity is produced when the following relationship is met

$$\tau_{oct} = a + b\sigma_{oct} \tag{7.3.2.1}$$

which is a linear expression that cuts the axes at points A and B shown in figure 7.3.2.1 and whose coordinates are

$$A\left(-\frac{a}{b},0\right) \quad B(0,a) \tag{7.3.2.2}$$

Equation (7.3.2.1) can be written according to the invariants of the stress fields. Substituting (3.8.1) and (3.8.2) into this expression gives

$$\sqrt{-6J_{2d}} = 3a + bJ_1 \tag{7.3.2.3}$$

where J_{2d} is the second invariant of the deviatoric tensor and J_1 is the first invariant of the complete stress field.

Equation (7.3.2.1), or the one indicated in (7.3.2.3), represents a plastification surface that has a conical shape and symmetry of revolution with

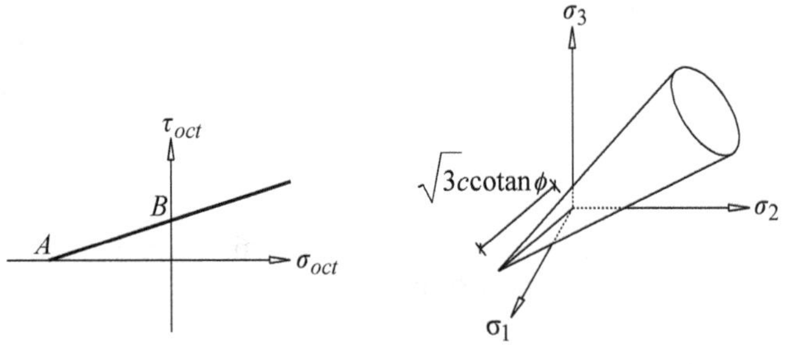

a) In (σ_{oct}, τ_{oct}) axes b) In (σ_1, σ_2, σ_3) axes

Figure 7.3.2.1: Graphic representation of the Drucker–Prager criterion regard to the trisector line in the coordinate system defined by the principal directions. This approach assumes that the cone has the vertex at the same point as the criterion of Mohr–Coulomb, and at this point

$$\sigma_{oct} = -c\text{cotan}\phi \quad \tau_{oct} = 0 \tag{7.3.2.4}$$

therefore

$$a = bc\text{cotan}\phi \tag{7.3.2.5}$$

Equation (7.3.2.1) then becomes

$$\tau_{oct} = bc\text{cotan}\phi + b\sigma_{oct} \tag{7.3.2.6}$$

The Drucker–Prager criterion can match that of the Mohr–Coulomb by fitting it to the vertices of higher or lower value of the latter plastification surface, depending on what is considered more appropriate for a given material, giving rise to the diagrams shown in figures 7.3.2.2.

The version of the Drucker–Prager criterion fitted to the vertices of higher value of the Mohr–Coulomb gives

$$b = \frac{2\sqrt{2}\sin\phi}{3 - \sin\phi} \tag{7.3.2.7}$$

and the expression for the criterion results in

$$\tau_{oct} = \frac{2\sqrt{2}c\cos\phi}{3 - \sin\phi} + \frac{2\sqrt{2}\sin\phi}{3 - \sin\phi}\sigma_{oct} \tag{7.3.2.8}$$

using the invariants of the stress fields given by (7.3.2.3) gives the following formula

$$\sqrt{-3J_{2d}} = \frac{6c\cos\phi}{3 - \sin\phi} + \frac{2\sin\phi}{3 - \sin\phi}J_1 \tag{7.3.2.9}$$

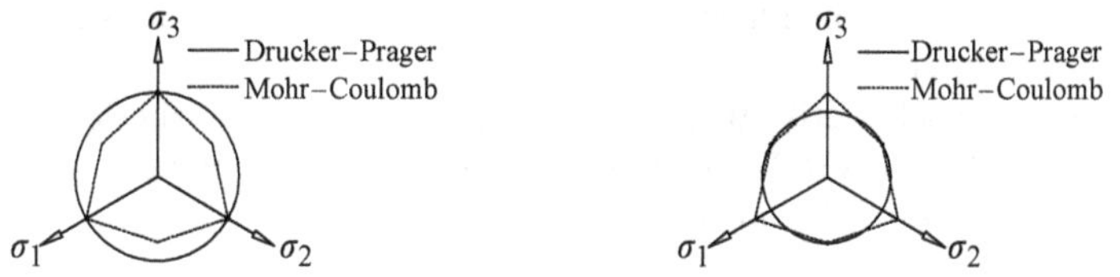

a) Fitting to the vertices of higher value b) Fitting to the vertices of lower value

Figure 7.3.2.2: Fittings between the Drucker–Prager and Mohr–Coulomb criteria

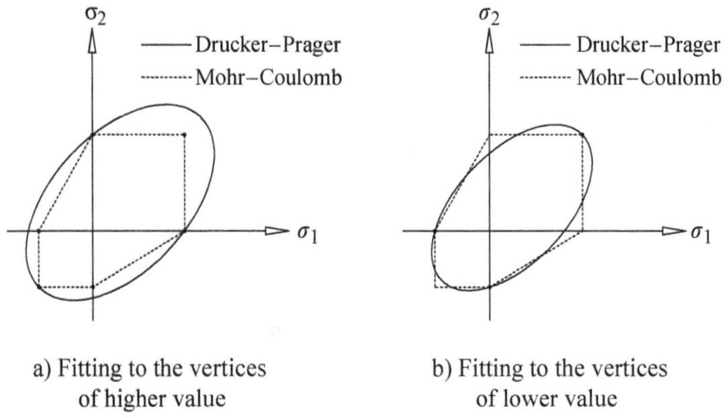

Figure 7.3.2.3: Sections in the $\sigma_3 = 0$ plane of the Drucker–Prager criterion

If the fitting is made to the vertices of lower value, the value of b is

$$b = \frac{2\sqrt{2}\sin\phi}{3+\sin\phi} \tag{7.3.2.10}$$

as a result, the equation of the criterion becomes

$$\tau_{oct} = \frac{2\sqrt{2}c\cos\phi}{3+\sin\phi} + \frac{2\sqrt{2}\sin\phi}{3+\sin\phi}\sigma_{oct} \tag{7.3.2.11}$$

which can be written as a function of the invariants as

$$\sqrt{-3J_{2d}} = \frac{6c\cos\phi}{3+\sin\phi} + \frac{2\sin\phi}{3+\sin\phi}J_1 \tag{7.3.2.12}$$

The sections on the plane defined by σ_1, σ_2 for the two cases produce the diagrams in figures 7.3.2.3.

One can easily verify that if $\phi = 0$ and $c = 0.5\sigma_e$, this criterion coincides with that of Von Mises–Hencky as equation (7.3.2.3) is the same as (7.2.2.9) and equations (7.3.2.8) and (7.3.2.11) are the same as (7.2.2.11).

7.3.3 Cambridge model. Cam Clay and modified Cam Clay

These plasticity criteria were set forth by researchers from the University of Cambridge (Roscoe and Burland) and are used in soil mechanics, mainly in clay materials.

Both criteria, Cam Clay (CC) and modified Cam Clay (MCC), are based on the same hypothesis, which establishes that the octahedral shear stress τ_{oct} and the octahedral normal stress σ_{oct} cannot exceed a particular proportion of each material. The line that relates the two is called the *critical state line*, and the expression is

$$\tau_{oct} = k\sigma_{oct} \tag{7.3.3.1}$$

This line is represented in figure 7.3.3.1.

The difference between the two models lies in the geometry of the *yield surface*. Its expression in each case is as follows

– Cam Clay model

$$\tau_{oct} + k\sigma_{oct} \ln\left(\frac{\sigma_{oct}}{\sigma_p}\right) = 0 \tag{7.3.3.2}$$

– Modified Cam Clay model

$$\tau_{oct}^2 - k^2\sigma_{oct}\left(\sigma_p - \sigma_{oct}\right) = 0 \tag{7.3.3.3}$$

In these formulas, σ_p is the *pre-consolidation stress*, which represents the maximum compressive stress that the material has experienced in the past. In figure 7.3.3.1, both yield surfaces appear together with the critical state line.

In both models, the maximum value of τ_{oct} on the yield surfaces coincides with the crossing point of the critical state line.

In these figures, points *A* represent an unstable situation of the soil because they are above the critical state line. In contrast, points *B* represent stable conditions. In figure 7.3.3.2, the three-dimensional view of the yield surface of the modified Cam Clay model and its representation in the Haigh–Westergaard space is shown. Similar figures can be obtained for the Cam Clay criterion.

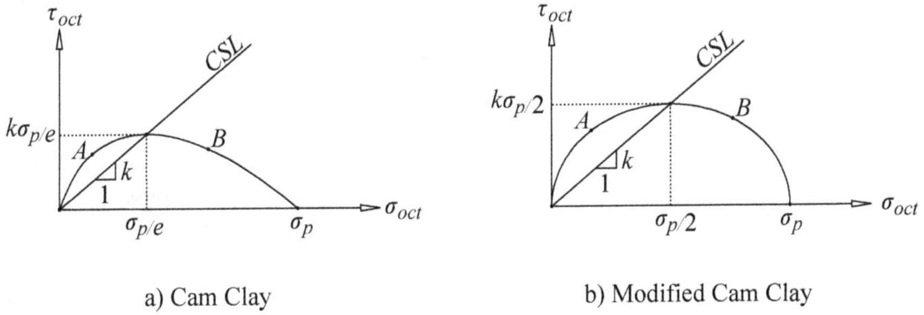

Figure 7.3.3.1: Cam Clay model and modified Cam Clay model

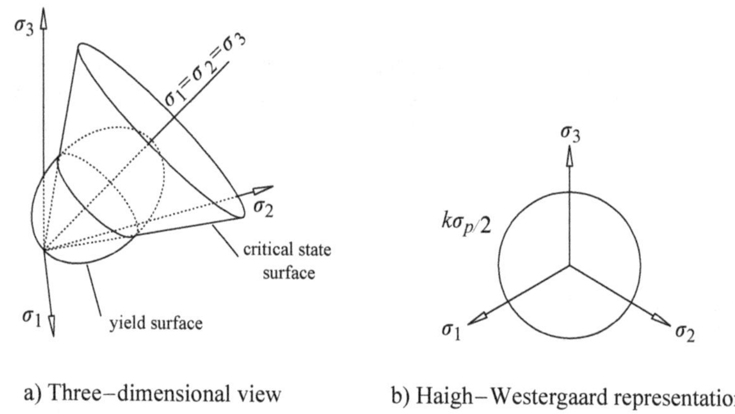

Figure 7.3.3.2: Modified Cam Clay criterion

The soil's behaviour when faced with new stress fields depends on its initial conditions, according to the CC and CCM models. Thus, one can distinguish the following situations:

a) Slightly pre-consolidated soil: this means that σ_{oct} is not much lower than σ_p.
b) Highly pre-consolidated soil: this means that $\sigma_{oct} \ll \sigma_p$.

The soil's behaviour in each of these fields according to the CCM model can be described as follows.

Suppose a first situation that starts from a stress value σ_A defined by point *A*, and then the stress increases gradually along a straight line, with the relationship between the normal and shear stress defined by $\tau = m\,(\sigma - \sigma_A)$. As shown in figure 7.3.3.3, the stress field will move from point *A* to point *B*, located on the yield surface. However, as the point is situated below the critical state line, the stress in the soil can continue to increase until it reaches point *C*, which is located on the critical state line. One can observe that, in this situation, the yield surface has increased its size and cuts the axis of abscissas at the point $2\tau_C/k$, which indicates that the phenomenon of strain hardening has occurred.

Suppose the opposite case, where a soil is very pre-consolidated, as shown in figure 7.3.3.4. If one starts from point *A* and the stresses increase following the ratio $\tau = m\,(\sigma - \sigma_A)$, one will arrives at point *B*, which is located on the yield surface. But as the point is above the critical state line, for this stress field the soil will enter into a plastic phase, and to avoid this, one will have to decrease the stresses, reducing them by following the previous line, but in the opposite direction until it reaches point *C*, located in the critical state line. In

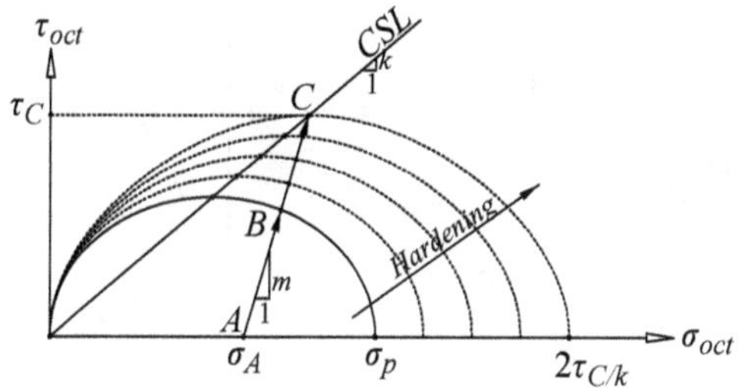

Figure 7.3.3.3: Strain hardening in slightly pre-consolidated soil

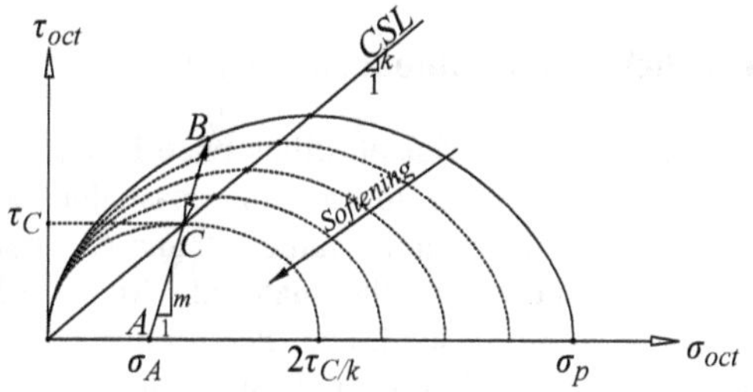

Figure 7.3.3.4: Strain softening in highly pre-consolidated soils

this case, the yield surface of the soil has been reduced with respect to its initial situation, and the new ellipse cuts the axis of abscissas at a distance $2\tau_C/k$ from the coordinates' origin, which means that strain softening has occurred.

In the case of the CC model, the phenomenon is identical, the only variation being the geometry of the yield surface.

7.4 ELASTOPLASTIC BEHAVIOUR OF MATERIALS

In the previous paragraphs, different plasticity criteria for materials were described, that is to say, stress fields that exhaust their strength capacity. Thereafter, any increase in the loads leads to an increment in the strain in the material without modifying the stress. However, it may happen that after the material enters into plasticity, for instance, with tensile stresses, an unloading cycle starts. In this case, part of the strain is recovered and it decreases following a line parallel to the loading phase. If the material similarly supports both compressive and tensile stresses, the unloading phase can be followed by another loading phase of compression, and this cycle can be repeated successively.

Two formulations for this type of behaviour will be described.

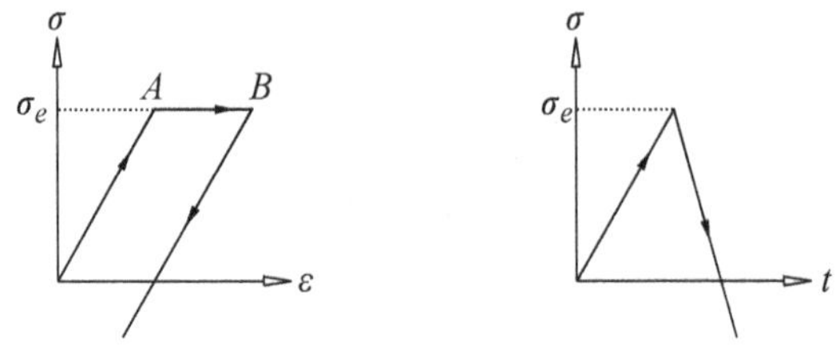

a) Loading and unloading cycle b) Time evolution of the load

Figure 7.4.1: Elastoplastic behaviour of a material

7.4.1 Elastoplastic behaviour without strain hardening

In this type of behaviour, a material plastifies once it reaches the maximum stress σ_e (figure 7.4.1.1). As a result, in a loading phase that terminates at point A, the stress and the strain will reach the values σ_e and $\varepsilon_e = \sigma_e/E$, respectively, E being the elasticity modulus of the material. Afterwards, the material plastifies without any increase in load, and the strain plastically increases its value ε_p until it reaches point B. With the unloading of the material, if the behaviour is symmetrical in tension and compression, the material will undergo a decrease of $2\varepsilon_e$ in the elastic strain until it reaches point D. The new plastic phase reduces the strain once more by an amount ε_p until it reaches point E. Once there, a final loading phase would return the trace to the coordinates' origin.

Table 7.4.1.1 gives a summary of the stress and strain values at each point of figure 7.4.1.1.

In successive loading/unloading cycles, the speeds at which the loads are applied or the material's plasticity times do not have to be identical. In this case, the previous trace does not pass through the coordinates' origin, as shown in figure 7.4.1.1b.

Table 7.4.1.1: Stress and strain values at each point

Point	**Stress**	**Strain**
A	σ_e	ε_e
B	σ_e	$\varepsilon_e + \varepsilon_p$
C	0	ε_p
D	$-\sigma_e$	$\varepsilon_p - \varepsilon_e$
E	$-\sigma_e$	$-\varepsilon_e$

To conceptually represent the elastoplastic behaviour, one normally uses the diagram shown in figure 7.4.1.2b, where, to a linear spring with stiffness E, a friction element is added that is activated when the stress reaches a value of σ_e. As a result, the model works as follows:

a) When $\sigma < \sigma_e$, the friction element remains fixed and all the strain corresponds to the linear spring

$$\varepsilon = \frac{\sigma}{E} \tag{7.4.1.1a}$$

b) When $\sigma = \sigma_e$, then, to the elastic strain $\varepsilon_e = \sigma_e/E$ (point A), a friction element is added, which results in an additional strain ε_p (point B); therefore

$$\varepsilon_B = \varepsilon_e + \varepsilon_p = \frac{\sigma_e}{E} + \varepsilon_p \tag{7.4.1.1b}$$

Stress fields are not possible when $\sigma > \sigma_e$.

c) If the unloading is initiated, the friction element stays where it was after finishing the plastic phase because the stress will be lower than σ_e; thus, for a stress $\sigma < \sigma_e$, the strain will be

$$\varepsilon = \varepsilon_p + \frac{\sigma}{E} \tag{7.4.1.1c}$$

When $\sigma = 0$, one obtains point C on the graph. From here, the spring may continue to deform due to compression until the stress is $\sigma = -\sigma_e$,

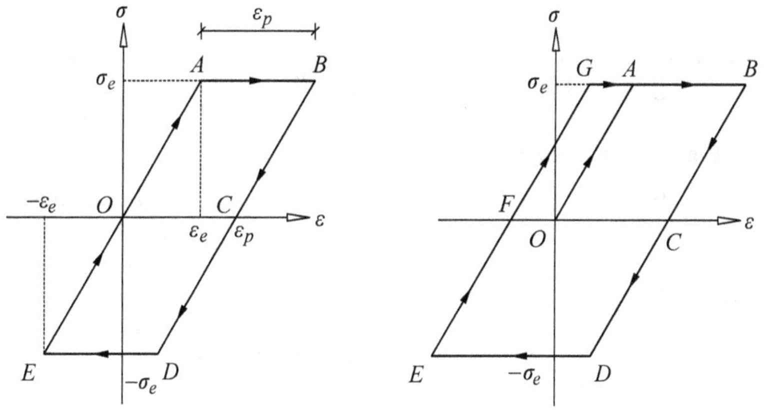

a) Elastoplastic cycle b) Elastoplastic cycle with hysteresis

Figure 7.4.1.1: Elastoplastic behaviour without hardening

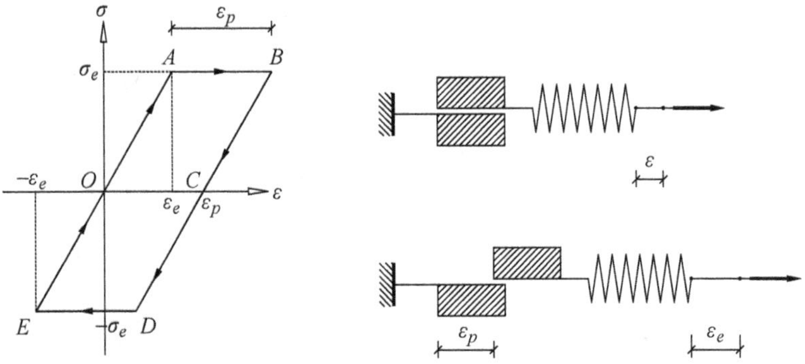

a) Elastoplastic cycle

b) Friction element and linear spring in elasticphase and in plastic phase

Figure 7.4.1.2: Elastoplastic behaviour without hardening

which corresponds to point *D*, for which

$$\varepsilon_D = \varepsilon_p - \varepsilon_e \tag{7.4.1.1d}$$

d) As the stress reaches σ_e, the friction element is activated and the linear spring does not support greater stress; thus, one comes to point *E* where

$$\varepsilon_E = -\varepsilon_e \tag{7.4.1.1e}$$

e) From this point, a new loading process produces the deformation of the spring element until it reaches the coordinates origin.

7.4.2 Elastoplastic behaviour with strain hardening

In this model, when the material reaches the elastic stress σ_e, plasticity is initiated, although a percentage of the initial elastic behaviour is maintained. This is described in figure 7.4.2.1a. When unloading the material, the elastic branch is longer; this produces the effect known as *strain hardening*.

One can observe that the model in figure 7.4.2.1b can be represented by a parallel assembly of a linear spring and a combination of a linear spring and a friction element that is activated for the stress value $E_1 \varepsilon_e$.

Consequently, when faced with a stress σ, the strain of each of the two parts of the model should be the same, and the stresses will be, respectively, σ_1 and σ_2.

The model's behaviour in each stage is:

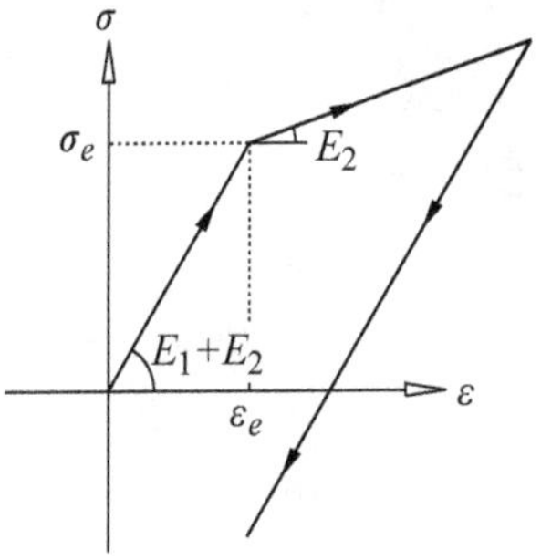

a) Elastoplastic behaviour with strain hardening

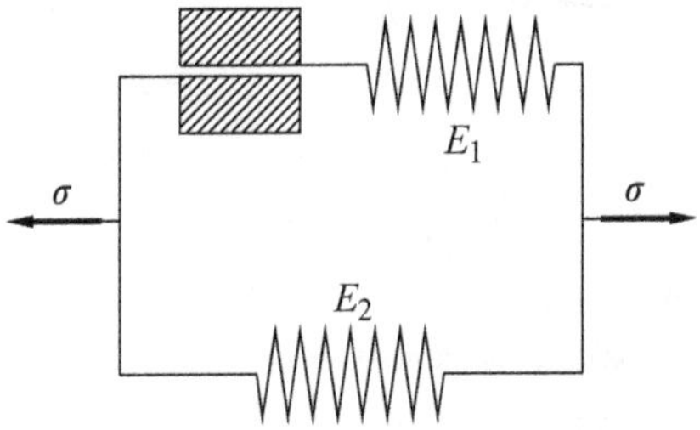

b) Model with a friction element and linear springs

Figure 7.4.2.1: Elastoplastic behaviour with strain hardening

a) Strain $\varepsilon < \varepsilon_e$. The friction element is not activated. The strain ε is the same in the two springs and gives

$$\sigma_1 = E_1\varepsilon \quad \sigma_2 = E_2\varepsilon \quad \sigma = \sigma_1 + \sigma_2 \tag{7.4.2.1a}$$

therefore

$$\sigma = (E_1 + E_2)\,\varepsilon \tag{7.4.2.1b}$$

This equation shows that the slope of the straight line is $E_1 + E_2$.

b) Strain $\varepsilon = \varepsilon_e$. In this situation, the stress on the spring with constant E_1 is the maximum admissible, and equation (7.4.2.1b) is transformed into

$$\sigma_e = (E_1 + E_2)\,\varepsilon_e \tag{7.4.2.2}$$

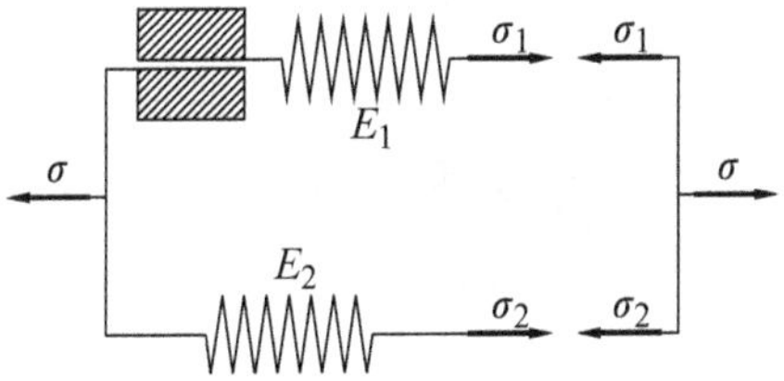

a) Stress diagram of the model

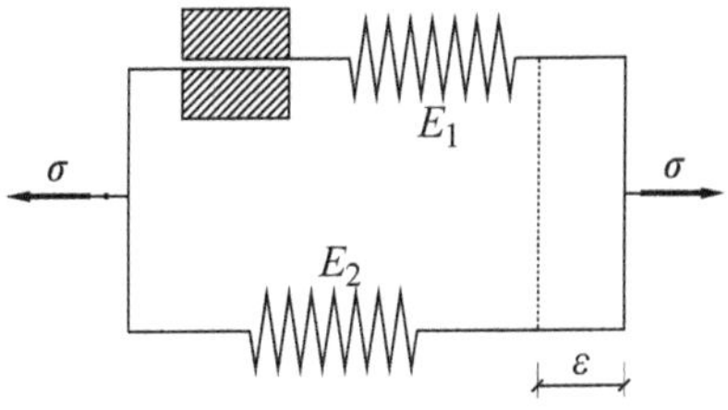

b) Strain diagram of the model

Figure 7.4.2.2: Stresses and strains in the elastoplastic model with strain hardening

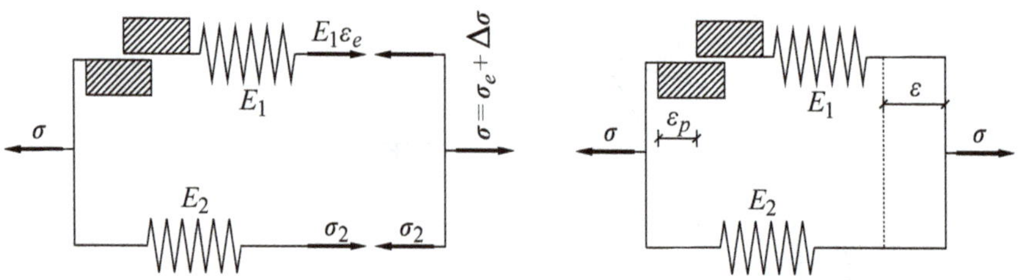

a) Stress diagram of the model b) Strain diagram of the model

Figure 7.4.2.3: Stresses and strains after activating the friction element

This equation implies that the friction element is activated and that the spring stiffness E_1 will not accept more stress.

c) Strain $\varepsilon > \varepsilon_e$. The strain continues being the same in both parts of the model and all the stress increase will be supported by the spring with constant E_2. This results in

$$\Delta\sigma = \sigma - \sigma_e = \Delta\sigma_2 = E_2\left(\varepsilon - \varepsilon_e\right) \qquad (7.4.2.3)$$

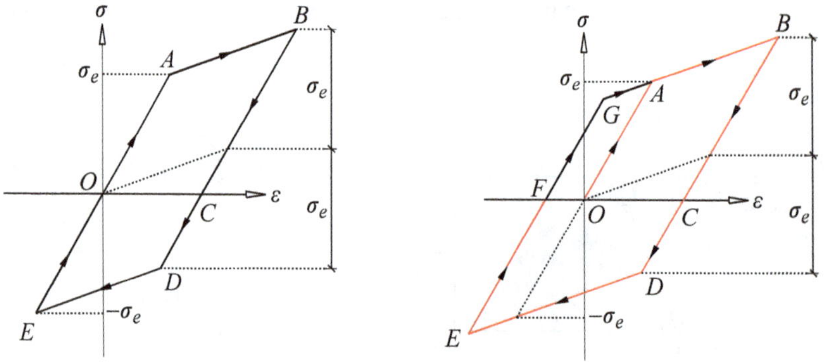

a) Elastoplastic cycle with strain hardening b) Elastoplastic cycle with strain hardening and hysteresis

Figure 7.4.2.4: Elastoplastic material with strain hardening

Equation (7.4.2.3) shows that the straight line that represents this section has slope E_2. In addition

$$\sigma = \sigma_e + \Delta\sigma = (E_1 + E_2)\,\varepsilon_e + E_2\,(\varepsilon - \varepsilon_e) = E_1\varepsilon_e + E_2\varepsilon \qquad (7.4.2.4)$$

$$\varepsilon - \varepsilon_e = \varepsilon_p \qquad (7.4.2.5)$$

These expressions show that the spring with stiffness E_2 is still in its elastic phase and, with stiffness E_1, cannot support any further stress. Furthermore, the differences between the strains ε and ε_e correspond to the plastic strain, as modelled by the friction element.

If the material is unloaded and afterwards loaded with compressive stress, the stages will be those that appear in figure 7.4.2.4. As previously, figure 7.4.2.4 shows that the cycle does not have to pass through the origin of coordinates, and in this case the graph from figure 7.4.2.4b applies.

EXERCISES

7.1. Given the stress field defined by the stress tensor

$$\tau = \begin{bmatrix} 3 & 1 & 1 \\ 1 & 0 & 2 \\ 1 & 2 & 0 \end{bmatrix}$$

and knowing that in this material $\nu = 0.3$ and $\sigma_e = 7$, find the linear increase in the stresses that will produce plasticity with the following criteria:

1. Beltrami–Haigh criterion.
2. Von Mises–Hencky criterion.
3. Tresca criterion.

7.2. For Mohr–Coulomb's plasticity criterion, find:

1. The distance from the vertex of the plastic surface to the coordinates' origin.
2. Equations for the section of the plastic surface in the plane defined by the principal stresses σ_1, σ_2.

7.3. For Drucker–Prager's plasticity criterion:

1. Fitting to the vertices of higher value of the Mohr–Coulomb criterion:

   a. Demonstrate the value of equation (7.3.2.7).
   b. Find the section in the plane defined by σ_1, σ_2.

c. Find the radius of the circle that corresponds to this criterion in the Haigh–Westergaard representation.

2. Fitting to the vertices of lower value of the Mohr–Coulomb criterion:

a. Demonstrate the value of equation (7.3.2.10).
b. Find the section in the plane defined by σ_1, σ_2.
c. Find the radius of the circle that corresponds to this criterion in the Haigh–Westergaard representation.

7.4. The behaviour of a clay material can be studied using the modified Cam Clay criterion. Its stress field is currently $\sigma_{oct} = 0.5\sigma_p$ and $\tau_{oct} = 0$. A new loading phase introduces new stresses in the material that will increase following a linear law of the form $\tau_{oct} = \left(\sigma_{oct} - 0.5\sigma_p\right)\tan\varphi$. Knowing the constant k of the critical state line and the angle φ, i.e.:

$$k = 1 \quad \varphi = 63.435°$$

1. Find the stress values at which the soil will no longer have an elastic behaviour.
2. Find the maximum values of the stresses that the material can withstand in accordance with the law.

7.5. The model that appears in the figure below has been proposed for the elastoplastic behaviour of a material. It is composed of a friction element, a spring in series and another spring in parallel. The friction element is activated for stress σ_e and the springs have stiffness E_1 and E_2. Establish the model's equations in each of the elastoplastic phases and the slope of the straight lines in the diagram (σ, ε).

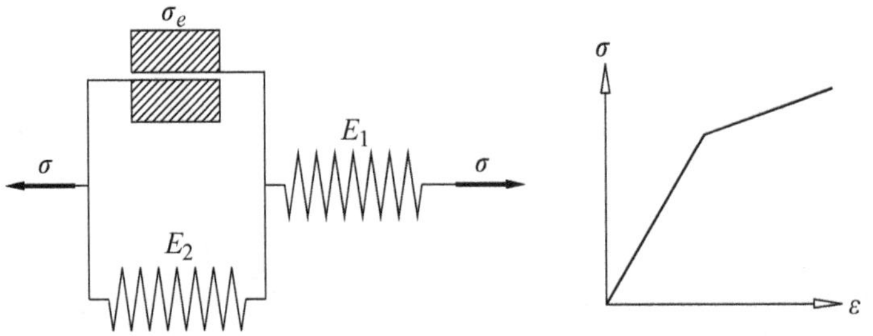

7.6. Assuming that the plate's material in exercise 5.7 obeys Mohr–Coulomb's plasticity criterion, defined by a cohesion c = 100 kPa and friction angle ϕ = 30°, calculate the value of a for which plasticity will be reached.

7.7. The following figures represent two cycles of hysteresis obtained from two specimen tests of specimens of a material with elastoplastic behaviour and strain hardening as defined by the model with friction element and linear springs.

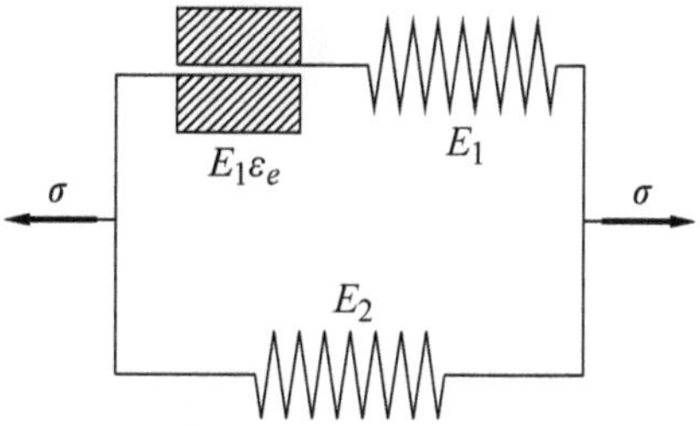

Data from some of the points in each test are as follows:

Trial I

Point A: $\sigma_e = 273$ MPa; $\varepsilon_e = 1.3\times10^{-3}$

Point B: $\sigma = 350$ MPa; $\varepsilon = 2.3\times10^{-3}$

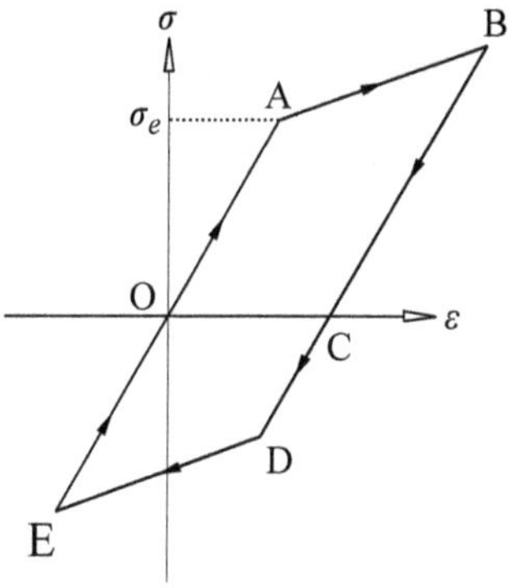

Trial II

Point A: $\sigma_e = 273$ MPa; $\varepsilon_e = 1.3\times10^{-3}$

Point B: $\sigma = 350$ MPa; $\varepsilon = 2.3\times10^{-3}$

Point E: $\varepsilon = -2.167\times10^{-3}$

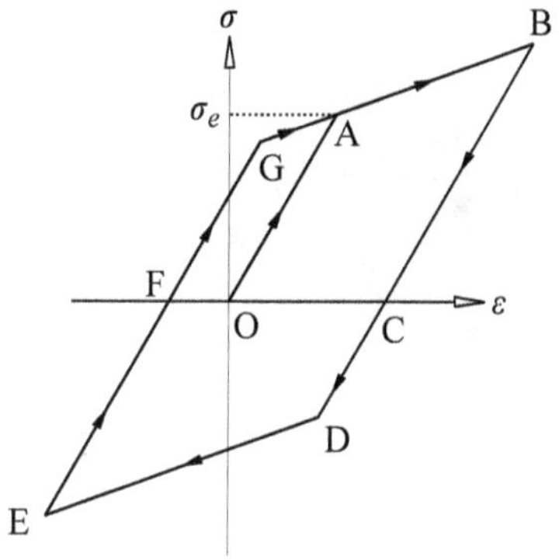

Find:

1. The values of E_1 and E_2.
2. The stress and strain values for each of the points that define the geometry of the hysteresis cycle in each test.

CHAPTER 8

LINEAR VISCOELASTICITY

8.1 INTRODUCTION

Some materials even within the stress range of elastic behaviour experience gradual strain increases without a change of stress.

As a consequence, when the load is applied, the elasticity modulus appears to be time-dependent. This phenomenon can be expressed as follows

$$\text{for } t = t_0 \qquad \sigma = \sigma_0 \qquad E = E_0 \qquad \varepsilon = \varepsilon_0 \tag{8.1.1a}$$

$$\sigma_0 = E(t)\,\varepsilon(t) \tag{8.1.1b}$$

$$E(t) < E(t_0) \qquad \varepsilon(t) > \varepsilon(t_0) \tag{8.1.1c}$$

This behaviour of the material is known as creep.

Other materials behave in the opposite manner. In other words, the deformations remain constant even with a decreasing stress, i.e.

$$\text{for } t = t_0 \qquad \sigma = \sigma_0 \qquad E = E_0 \qquad \varepsilon = \varepsilon_0 \tag{8.1.2a}$$

$$\sigma(t) = E(t)\,\varepsilon_0 \tag{8.1.2b}$$

$$\sigma(t) < \sigma(t_0) \qquad E(t) < E(t_0) \tag{8.1.2c}$$

As one can observe, the stresses in the material decrease with time, and for this reason, this behaviour is known as *relaxation.*

Both phenomena have the common denomination of *viscoelasticity*. There are several models to study the viscoelastic behaviour of materials, and the most significant are presented in the following sections.

8.2 MODELS OF VISCOELASTIC BEHAVIOUR

There are two types of elements in all the viscoelastic models. One of them is represented by an elastic and linear spring whose equation is

$$\sigma = E\varepsilon \tag{8.2.1}$$

This element simulates the initial strain that the load produces on the material, and it is a behaviour similar to that of elastic solids.

Figure 8.2.1: Elements of the viscoelastic models

The other element is a viscous linear damper defined by the damping constant η, which relates stress and strain in the following way

$$\sigma = \eta \dot{\varepsilon} \tag{8.2.2}$$

This element simulates the time-dependent strain that is produced, and it is similar to that of viscous fluids. Therefore, a viscoelastic medium combines both solid and fluid behaviour.

Using these two elements, several models of increasing complexity can be defined.

8.2.1 Two-parameter models. Maxwell and Voigt models

Maxwell's model consists of combining in series the spring and the linear damper, as shown in figure 8.2.1.1.

The equilibrium and strain equations for this model are

$$\sigma = \sigma_E = \sigma_A \tag{8.2.1.1}$$

$$\varepsilon = \varepsilon_E + \varepsilon_A \tag{8.2.1.2}$$

where σ, σ_E, σ_A are the total stress in the material, the stress in the spring and the stress in the damper, all of which are equal, and ε, ε_E, ε_A are the total strain and the components of the strain in the spring and the damper, respectively.

To obtain the strain over time in a body subjected to a constant stress σ_0 that will produce an initial strain ε_0, one just has to consider the derivative of equation (8.2.1.2) with respect to time. The strain ε_E is constant, because it depends on the stress, which is also constant, and equal to the total initial strain, that is to say, $\varepsilon_E = \varepsilon_0$. Therefore, one has

$$\dot{\varepsilon} = \dot{\varepsilon}_A \tag{8.2.1.3}$$

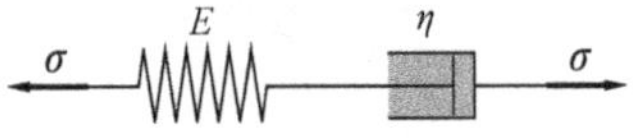

Figure 8.2.1.1: Maxwell's viscoelastic model

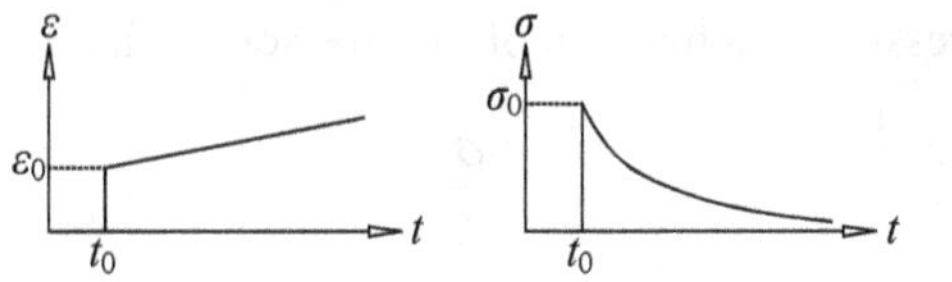

Figure 8.2.1.2: Creep and relaxation functions from Maxwell's model

and

$$\dot{\varepsilon} = \frac{\sigma_0}{\eta} = \frac{E\varepsilon_0}{\eta} \tag{8.2.1.4}$$

which leads to

$$d\varepsilon = \frac{E\varepsilon_0}{\eta}dt \tag{8.2.1.5}$$

therefore

$$\varepsilon = \frac{E\varepsilon_0}{\eta}t + c \tag{8.2.1.6}$$

Imposing the initial conditions at $t = t_0$

$$\varepsilon_0 = \frac{E\varepsilon_0}{\eta}t_0 + c \tag{8.2.1.7}$$

leads to

$$c = \varepsilon_0\left(1 - \frac{Et_0}{\eta}\right) \tag{8.2.1.8}$$

and finally

$$\varepsilon = \varepsilon_0\left[1 + \frac{E}{\eta}(t - t_0)\right] \tag{8.2.1.9}$$

Equation (8.2.1.9) shows that the initial strain corresponds to that of the elastic spring and that the presence of the load increases the value of ε linearly, resembling the behaviour of fluids rather than solids.

If one wants to obtain the evolution of a stress in a solid that is subjected to a constant strain ε_0, with an initial stress σ_0, and thus the *relaxation function*, one needs to use the equation that defines the strains, i.e.

$$\varepsilon = \varepsilon_0 = \varepsilon_E + \varepsilon_A \tag{8.2.1.10}$$

differentiating with respect to time and remembering that the total strain must remain constant gives

$$0 = \dot{\varepsilon}_E + \dot{\varepsilon}_A \tag{8.2.1.11}$$

writing this expression as a function of the stresses results in

$$\frac{\dot{\sigma}}{E} + \frac{\sigma}{\eta} = 0 \tag{8.2.1.12}$$

or

$$\frac{d\sigma}{\sigma} = -\frac{E}{\eta}dt \tag{8.2.1.13}$$

therefore

$$\ln \sigma = -\frac{E}{\eta}t + c \tag{8.2.1.14}$$

which can also be written as

$$\sigma = e^{-\frac{E}{\eta}t + c} \tag{8.2.1.15}$$

Imposing the initial conditions at $t = t_0$

$$\sigma_0 = e^{-\frac{E}{\eta}t_0 + c} \tag{8.2.1.16}$$

which results in

$$e^c = \sigma_0 e^{\frac{E}{\eta}t_0} \tag{8.2.1.17}$$

by introducing this value into (8.2.1.15), one obtains the relaxation function, which is a decreasing function that tends towards zero.

$$\sigma = \sigma_0 e^{-\frac{E}{\eta}(t-t_0)} \tag{8.2.1.18}$$

This model does not properly reflect the behaviour of some of the most common materials, as the creep grows faster in concrete than as represented by a linear function. It also has a finite final value.

Voigt's model combines in parallel a viscous damper and an elastic spring, as shown in figure 8.2.1.3.

The equations that define this model are

$$\varepsilon = \varepsilon_E = \varepsilon_A \tag{8.2.1.19a}$$

$$\sigma = \sigma_E + \sigma_A = E\varepsilon_E + \eta\dot{\varepsilon}_A = E\varepsilon + \eta\dot{\varepsilon} \tag{8.2.1.19b}$$

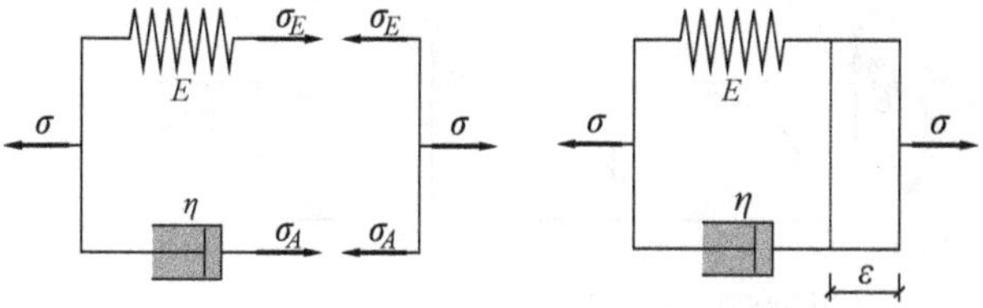

a) Stress diagram b) Strain diagram

Figure 8.2.1.3: Voigt's viscoelastic model

To obtain the function that describes the creep mechanism caused by a constant stress σ_0, one uses the first and last terms of the equilibrium equation, i.e.

$$\eta\dot{\varepsilon} = \sigma_0 - E\varepsilon \tag{8.2.1.20}$$

which can be also written as

$$\frac{d\varepsilon}{\varepsilon - \dfrac{\sigma_0}{E}} = -\frac{E}{\eta}dt \tag{8.2.1.21}$$

performing the integration gives

$$\ln\left(\varepsilon - \frac{\sigma_0}{E}\right) = -\frac{E}{\eta}t + c \tag{8.2.1.22}$$

or

$$\varepsilon - \frac{\sigma_0}{E} = e^{-\frac{E}{\eta}t + c} \tag{8.2.1.23}$$

The coefficient c is obtained by applying the initial conditions. It is assumed that, for $t = t_0$ $\varepsilon = \varepsilon_0 = 0$, there is no strain, which is equivalent to saying that at the beginning the force is absorbed only by the damper; therefore, for $t = t_0$

$$\sigma_E\left(t = t_0\right) = 0 \qquad \sigma_A\left(t = t_0\right) = \sigma \tag{8.2.1.24}$$

consequently

$$e^c = -\frac{\sigma_0}{E}e^{\frac{E}{\eta}t_0} \tag{8.2.1.25}$$

and, finally

$$\varepsilon = \frac{\sigma_0}{E}\left[1 - e^{-\frac{E}{\eta}(t-t_0)}\right] \tag{8.2.1.26}$$

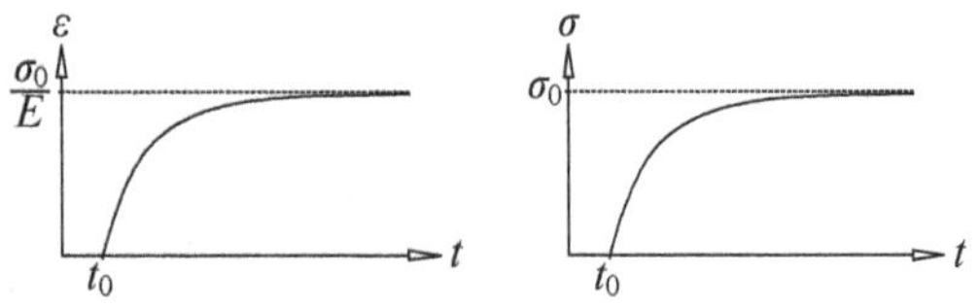

a) Creep function b) Evolution of stress in the elastic spring

Figure 8.2.1.4: Evolution of strain and stress in the Voigt model

One can conclude from the figures that the final strain is limited and it corresponds to that of the linear spring under the stress σ_0. That is to say, the damper takes the force at the beginning and the spring at the end of the process. In addition, as the strain is limited, this behaviour is similar to that of solids. A disadvantage of this model is that the initial strain is zero, which does not reflect the behaviour of any material.

The relaxation phenomenon is not well represented by Voigt's model, and for this reason a description of it is not included.

8.2.2 Three-parameter models. Maxwell's and Voigt's standard linear models

A formulation that adds another element to the two that were present in the previous models was introduced by Zener, giving place to three-parameter models, the most common being those known as SLS (*standard linear solid*) by Maxwell and Voigt. The first combines in parallel an element composed of an elastic spring with stiffness E_1 and a linear viscous damper with constant η with another elastic spring of stiffness E_2, as shown in figure 8.2.2.1a.

Voigt's SLS model contains a Voigt's element and an elastic spring, combined in series as described in figure 8.2.2.1b.

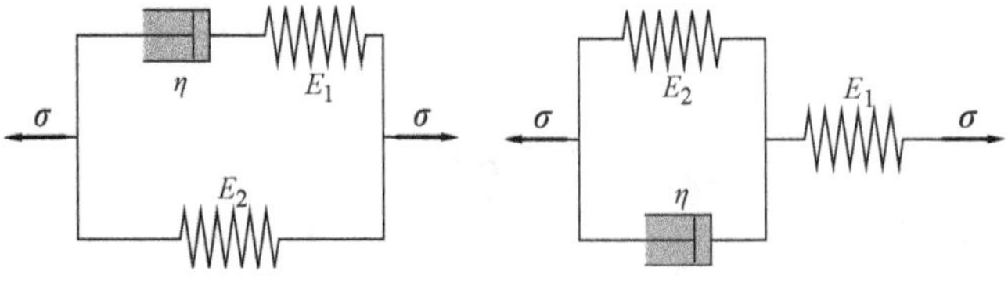

a) Maxwell's SLS model b) Voigt's SLS model

Figure 8.2.2.1: Viscoelastic SLS models

The stress and strain equations that define Maxwell's SLS model are

$$\varepsilon = \varepsilon_1 + \varepsilon_A = \varepsilon_2 \tag{8.2.2.1a}$$

$$\sigma_1 = E_1\varepsilon_1 = \eta\dot{\varepsilon}_A \qquad \sigma_2 = E_2\varepsilon_2 = E_2\varepsilon \tag{8.2.2.1b}$$

$$\sigma = \sigma_1 + \sigma_2 \tag{8.2.2.1c}$$

$$\sigma_0 = (E_1 + E_2)\,\varepsilon_0 = E_0\varepsilon_0 \tag{8.2.2.1d}$$

where ε is the total strain, which coincides with the strain of the spring with stiffness E_2; ε_1 and ε_A are the strains of the spring with stiffness E_1 and the viscous damper; $\dot{\varepsilon}_A$ is the speed of the strain of that element; ε_0 and σ_0 are the initial strain and stress; E_0 is the initial modulus of elasticity of the material, and σ, σ_1 and σ_2 are the total stress and the components of Maxwell's model and of the spring with stiffness E_2.

The function that describes the creep mechanism, i.e. defines the strain produced by a constant stress, is obtained by differentiating the equation of compatibility of strains established in (8.2.2.1a), i.e.

$$\dot{\varepsilon} = \dot{\varepsilon}_1 + \dot{\varepsilon}_A = \frac{\dot{\sigma}_1}{E_1} + \frac{\sigma_1}{\eta} \tag{8.2.2.2}$$

The stress equilibrium gives

$$\dot{\varepsilon} = \frac{\dot{\sigma} - \dot{\sigma}_2}{E_1} + \frac{\sigma - \sigma_2}{\eta} \tag{8.2.2.3}$$

as the stress is constant, $\dot{\sigma}$ is equal to zero. By expressing the stresses as a function of the strains, one obtains

$$\dot{\varepsilon} = \frac{-E_2\dot{\varepsilon}}{E_1} + \frac{(E_1 + E_2)\,\varepsilon_0}{\eta} - \frac{E_2\varepsilon}{\eta} \tag{8.2.2.4}$$

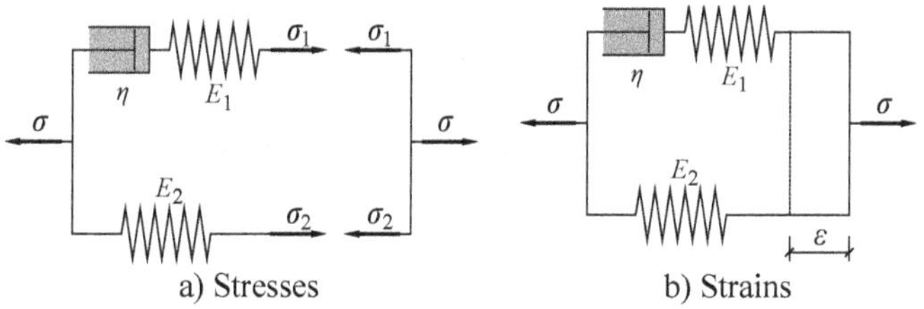

Figure 8.2.2.2: Stresses and strains in Maxwell's SLS viscoelastic model

and

$$\frac{\dot{\varepsilon}\,(E_1 + E_2)}{E_1} = \frac{(E_1 + E_2)\,\varepsilon_0}{\eta} - \frac{E_2 \varepsilon}{\eta} \tag{8.2.2.5}$$

rearranging the expression

$$\frac{\dot{\varepsilon}}{\dfrac{E_2\varepsilon - (E_1 + E_2)\,\varepsilon_0}{E_1 + E_2}} = -\frac{E_1}{\eta} \tag{8.2.2.6}$$

which can be written as

$$\frac{d\varepsilon}{\dfrac{E_2\varepsilon - (E_1 + E_2)\,\varepsilon_0}{E_1 + E_2}} = -\frac{E_1}{\eta}dt \tag{8.2.2.7}$$

performing the integrations, one obtains

$$\frac{E_1 + E_2}{E_2}\ln\left[\frac{E_2\varepsilon - (E_1 + E_2)\,\varepsilon_0}{E_1 + E_2}\right] = -\frac{E_1}{\eta}t + c \tag{8.2.2.8}$$

which can be written as

$$\frac{E_2\varepsilon - (E_1 + E_2)\,\varepsilon_0}{E_1 + E_2} = e^{-\frac{E_1 E_2}{(E_1 + E_2)\,\eta}t + c} \tag{8.2.2.9}$$

The initial conditions are $t = t_0$ and $\varepsilon = \varepsilon_0$, which gives

$$\frac{-E_1\varepsilon_0}{E_1 + E_2} = e^{-\frac{E_1 E_2}{(E_1 + E_2)\,\eta}t_0} e^c \tag{8.2.2.10}$$

and therefore

$$e^c = \frac{-E_1\varepsilon_0}{E_1 + E_2} e^{\frac{E_1 E_2}{(E_1 + E_2)\,\eta}t_0} \tag{8.2.2.11}$$

substituting into equation (8.2.2.9), one finally obtains

$$\varepsilon = \varepsilon_0\left[1 + \frac{E_1}{E_2}\left(1 - e^{-\frac{E_1 E_2}{(E_1 + E_2)\,\eta}(t - t_0)}\right)\right] \tag{8.2.2.12}$$

which is the expression for the creep function, which is represented in figure 8.2.2.3.

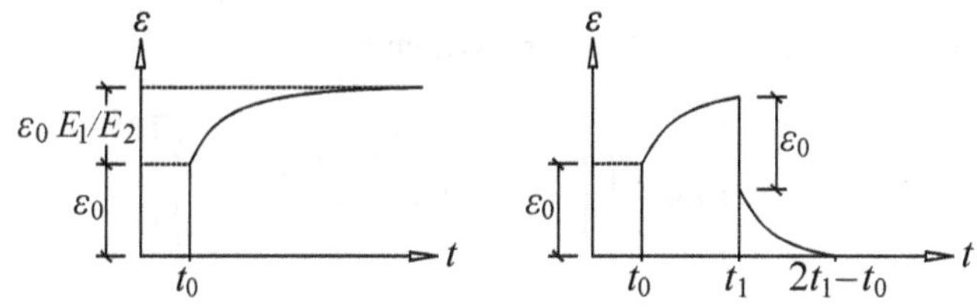

Figure 8.2.2.3: Creep function of Maxwell's SLS viscoelastic model

One can observe that the strain begins with the initial value ε_0, and afterwards a time-dependent term named *creep strain* is added. This term has a maximum value $\varepsilon_0 \cdot E_1/E_2$, which depends on each material. Figure 8.2.2.3a is a graphical representation of the creep function.

If the load acts during a limited time, the material's behaviour would be the one shown in figure 8.2.2.3b. The strain, after the initial value ε_0, will gradually increase and, at $t = t_1$, it will be that which appears in the figure and is expressed by

$$\varepsilon_{t_1} = \varepsilon_0 \left[1 + \frac{E_1}{E_2} \left(1 - e^{-\frac{E_1 E_2}{(E_1 + E_2)\,\eta}(t_1 - t_0)} \right) \right] \tag{8.2.2.13}$$

If the load then disappears, the instantaneous strain is recovered; thus, there will be a decrease in the value of ε_0, and afterwards the strain will undergo a continuous decrease defined by the following expression

$$\begin{aligned} \varepsilon &= \varepsilon_{t_1} - \varepsilon_0 \left[1 + \frac{E_1}{E_2} \left(1 - e^{-\frac{E_1 E_2}{(E_1 + E_2)\,\eta}(t - t_1)} \right) \right] \\ &= \varepsilon_0 \frac{E_1}{E_2} \left[e^{-\frac{E_1 E_2}{(E_1 + E_2)\,\eta}(t - t_1)} - e^{-\frac{E_1 E_2}{(E_1 + E_2)\,\eta}(t_1 - t_0)} \right] \end{aligned} \tag{8.2.2.14}$$

By removing the load at $t = t_1$, the strain is reduced to the value ε_0, which is the elastic part, and for $t = 2t_1 - t_0$ the strain completely disappears.

It is interesting to observe how each element of the SLS model participates throughout the process, and this can be done by observing the behaviour of the stresses σ_1 and σ_2 over time. They are written as

$$\sigma_2 = E_2 \varepsilon = E_2 \varepsilon_0 \left[1 + \frac{E_1}{E_2} \left(1 - e^{-\frac{E_1 E_2}{(E_1 + E_2)\eta}(t - t_0)} \right) \right] \tag{8.2.2.15}$$

$$\sigma_1 = \sigma - \sigma_2 = (E_1 + E_2)\varepsilon_0 - E_2 \varepsilon_0 \left[1 + \frac{E_1}{E_2} \left(1 - e^{-\frac{E_1 E_2}{(E_1 + E_2)\eta}(t - t_0)} \right) \right] \tag{8.2.2.16}$$

Equation (8.2.2.15) is the same as equation (8.2.2.12) multiplied by the elasticity modulus E_2, and its graphic representation is shown in figure 8.2.2.3a. The stress expression σ_1 finally becomes

$$\sigma_1 = E_1 \varepsilon_0 e^{-\frac{E_1 E_2}{(E_1 + E_2)\eta}(t - t_0)} \tag{8.2.2.17}$$

This stress has a maximum value when the load is applied, and it decreases monotonically until asymptotically reaching zero. As this stress is associated with the strain of the linear spring with stiffness E_1, equation (8.2.2.17) describes how the strain of this spring decreases over time until cancelling out.

On the other hand, the strain of the viscous damper will increase, although at a diminishing rate, so as to compensate the decrease in strain of the spring with stiffness E_1 and the increase in strain of the spring with stiffness E_2. Figure 8.2.2.4 shows the graphical representation of both stresses.

In this type of material, it is interesting to define the elasticity modulus $E(t) = E_t$, which relates the stress to the strain at each moment, and in particular its final value, i.e.

$$\sigma = E_t \varepsilon_t \qquad \sigma = E_\infty \varepsilon_\infty \tag{8.2.2.18}$$

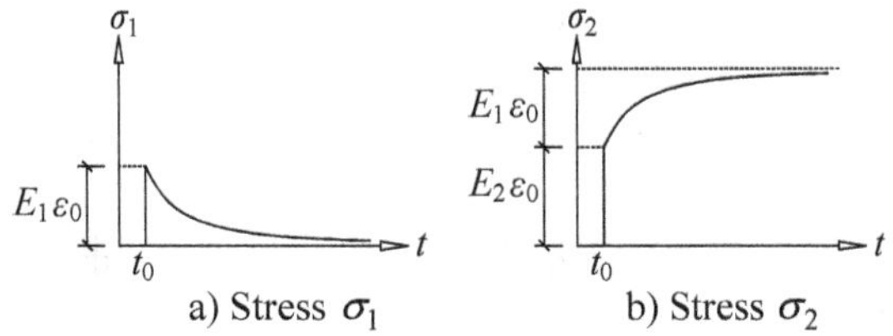

a) Stress σ_1 b) Stress σ_2

Figure 8.2.2.4: Evolution of the stresses σ_1 and σ_2 in Maxwell's SLS model

The second expression can be written

$$\sigma = E_\infty \varepsilon_\infty = E_\infty \left[\varepsilon_0 \left(1 + \frac{E_1}{E_2} \right) \right] \tag{8.2.2.19}$$

recalling (8.2.2.1d)

$$(E_1 + E_2)\, \varepsilon_0 = E_\infty \varepsilon_0 \left(1 + \frac{E_1}{E_2} \right) \tag{8.2.2.20}$$

it can then be seen that

$$E_\infty = E_2 \tag{8.2.2.21}$$

therefore, the elasticity modulus of the viscoelastic material at infinity is obtained by eliminating the spring's stiffness from Maxwell's element. The quantitative relevance of this decrease in stiffness can be calculated from the creep strain in the material. If one calls c_f the *creep coefficient*, the relationship between the strain produced by the creep, and the instantaneous strain, gives

$$c_f = \frac{\frac{E_1}{E_2}\varepsilon_0}{\varepsilon_0} = \frac{E_1}{E_2} \tag{8.2.2.22}$$

This formula represents the relationship between the stiffnesses of both springs. If the material initially has an elasticity modulus E_0, the stiffness values of both springs are

$$E_0 = E_1 + E_2 = E_2 \left(1 + c_f \right) \tag{8.2.2.23}$$

and therefore

$$E_1 = \frac{c_f}{1 + c_f} E_0 \qquad E_2 = \frac{E_0}{1 + c_f} \tag{8.2.2.24}$$

To obtain the relaxation function from an initial stress σ_0 that produces a constant strain ε_0, one similarly proceeds as follows

$$\varepsilon_0 = \varepsilon_1 + \varepsilon_A \tag{8.2.2.25}$$

calculating the derivative with respect to time and knowing that the strain remains constant gives

$$0 = \frac{\dot{\sigma}_1}{E_1} + \frac{\sigma_1}{\eta} = \frac{\dot{\sigma} - \dot{\sigma}_2}{E_1} + \frac{\sigma - \sigma_2}{\eta} \tag{8.2.2.26}$$

as the strain is constant, $\dot{\sigma}_2 = 0$ and $\sigma_2 = E_2\varepsilon_0$, then

$$\frac{\dot{\sigma}}{E_1} + \frac{\sigma - E_2\varepsilon_0}{\eta} = 0 \tag{8.2.2.27}$$

therefore

$$\frac{d\sigma}{\sigma - E_2\varepsilon_0} = -\frac{E_1}{\eta}dt \tag{8.2.2.28}$$

performing the integration gives

$$\ln\left(\sigma - E_2\varepsilon_0\right) = -\frac{E_1}{\eta}t + c \tag{8.2.2.29}$$

or

$$\sigma - E_2\varepsilon_0 = e^{-\frac{E_1}{\eta}t + c} \tag{8.2.2.30}$$

For $t = t_0$, one obtains

$$\left(E_1 + E_2\right)\varepsilon_0 - E_2\varepsilon_0 = e^{-\frac{E_1}{\eta}t_0} e^c \tag{8.2.2.31}$$

which leads to

$$e^c = E_1\varepsilon_0 e^{\frac{E_1}{\eta}t_0} \tag{8.2.2.32}$$

introducing this expression into (8.2.2.30) and recalling that $\sigma_0 = \left(E_1 + E_2\right)\varepsilon_0$, one finally obtains

$$\sigma = \sigma_0\left[1 - \frac{E_1}{E_1 + E_2}\left(1 - e^{-\frac{E_1}{\eta}(t-t_0)}\right)\right] = \sigma_0\frac{E_2}{E_1 + E_2}\left[1 + \frac{E_1}{E_2}e^{-\frac{E_1}{\eta}(t-t_0)}\right] \tag{8.2.2.33}$$

As time goes to infinity, the stress becomes

$$\sigma_\infty = \frac{E_2}{E_1 + E_2}\sigma_0 \tag{8.2.2.34}$$

therefore, the stress loss due to relaxation is given by

$$\Delta\sigma = \sigma_0 - \sigma_\infty = \left(E_1 + E_2\right)\varepsilon_0 - \frac{E_2\sigma_0}{E_1 + E_2} = \frac{E_1}{E_1 + E_2}\sigma_0 \tag{8.2.2.35}$$

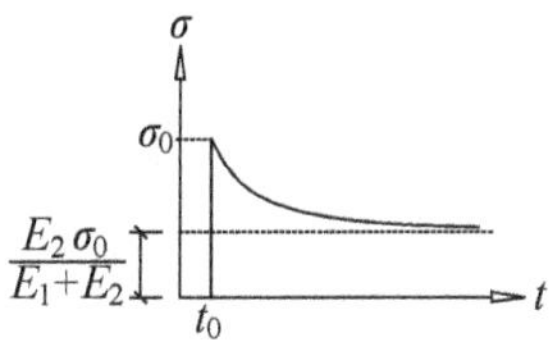

Figure 8.2.2.5: Relaxation function of Maxwell's SLS viscoelastic model

The stress loss percentage due to relaxation, or the *relaxation coefficient* c_r, is

$$c_r = \frac{E_1}{E_1 + E_2} \tag{8.2.2.36}$$

It is also interesting to obtain the expressions for the elasticity moduli E_1 and E_2 as a function of the relaxation coefficient c_r. One already knows from (8.2.2.1d) that

$$E_0 = E_1 + E_2 \tag{8.2.2.37}$$

which results in

$$E_1 = E_0 c_r \qquad E_2 = E_0 \left(1 - c_r\right) \tag{8.2.2.38}$$

taking into consideration that

$$\sigma_\infty = E_\infty \varepsilon_0 = \frac{E_2}{E_1 + E_2} \sigma_0 = E_2 \varepsilon_0 \tag{8.2.2.39}$$

the elasticity modulus at infinity E_∞ becomes

$$E_\infty = E_2 \tag{8.2.2.40}$$

8.2.3 Generalized standard linear models. Maxwell and Voigt models

The previous approach can be generalized if one considers an elastic spring and several of Maxwell's models in parallel, as shown in figure 8.2.3.1

In this model, E_E is the stiffness of the linear spring and E_i and η_i $(i = 1, \ldots, n)$ are the springs' stiffnesses and the dampers of the elements of Maxwell's models. The constitutive equations of the formulation are given by

$$\varepsilon = \varepsilon_E = \varepsilon_{Mi} + \varepsilon_{Ai} \quad i = 1, \ldots, n \tag{8.2.3.1a}$$

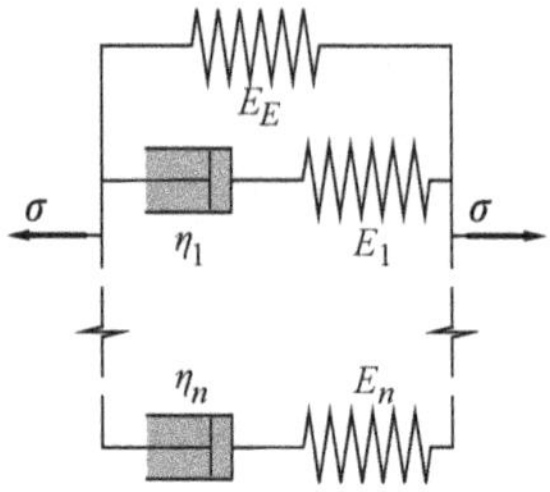

Figure 8.2.3.1: Maxwell's generalized SLS model

where ε_E is the strain of the linear spring and ε_{Mi} and ε_{Ai} are the strains of the springs and the dampers of the elements of Maxwell's models.

$$\sigma = \sigma_E + \sum_{i=1}^{n} \sigma_{Mi} \tag{8.2.3.1b}$$

$$\sigma_i = \sigma_{Mi} = \sigma_{Ai} = E_i \varepsilon_{Mi} = \eta_i \dot{\varepsilon}_{Ai} \quad i = 1, \ldots, n \tag{8.2.3.1c}$$

The relaxation function associated with a constant strain ε_0 produced by an initial stress σ_0 can be obtained from the equation

$$E_i \varepsilon_{Mi} = E_i (\varepsilon_0 - \varepsilon_{Ai}) = \eta_i \dot{\varepsilon}_{Ai} \quad i = 1, \ldots, n \tag{8.2.3.2}$$

at the beginning it complies with

$$\sigma_0 = E_E \varepsilon_0 + \sum_{i=1}^{n} E_i \varepsilon_0 = \left(E_E + \sum_{i=1}^{n} E_i \right) \varepsilon_0 \tag{8.2.3.3}$$

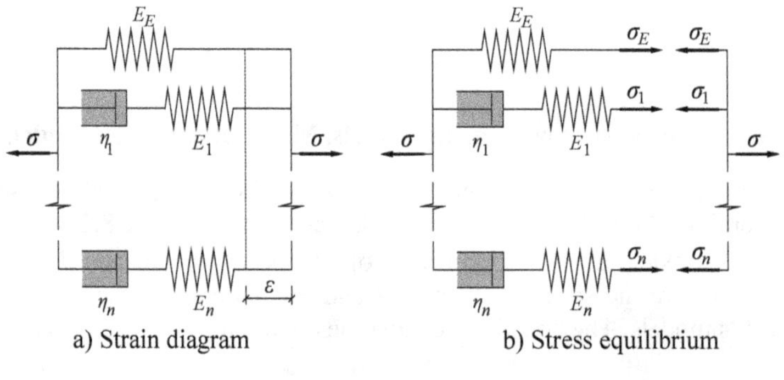

a) Strain diagram b) Stress equilibrium

Figure 8.2.3.2: Stress and strain in Maxwell's generalized SLS model one can define the total stiffness E_T of all the springs as

$$E_T = E_E + \sum_{i=1}^{n} E_i \tag{8.2.3.4}$$

and therefore

$$\varepsilon_0 = \frac{\sigma_0}{E_T} \tag{8.2.3.5}$$

Substituting the above expression into (8.2.3.2), one obtains

$$\dot{\varepsilon}_{Ai} = \frac{E_i}{\eta_i}\left(\frac{\sigma_0}{E_T} - \varepsilon_{Ai}\right) \quad i = 1,\ldots,n \tag{8.2.3.6}$$

operating on the above, one finds that

$$\frac{\dot{\varepsilon}_{Ai}}{\varepsilon_{Ai} - \dfrac{\sigma_0}{E_T}} = -\frac{E_i}{\eta_i} \quad i = 1,\ldots,n \tag{8.2.3.7}$$

and therefore

$$\frac{d\varepsilon_{Ai}}{\varepsilon_{Ai} - \dfrac{\sigma_0}{E_T}} = -\frac{E_i}{\eta_i}dt \quad i = 1,\ldots,n \tag{8.2.3.8}$$

performing the integration gives

$$\ln\left(\varepsilon_{Ai} - \frac{\sigma_0}{E_T}\right) = -\frac{E_i}{\eta_i}t + c \quad i = 1,\ldots,n \tag{8.2.3.9}$$

or

$$\varepsilon_{Ai} - \frac{\sigma_0}{E_T} = e^{-\frac{E_i}{\eta_i}t + c} \quad i = 1,\ldots,n \tag{8.2.3.10}$$

At the beginning of the process, $t = t_0$ and $\varepsilon_{Ai} = 0$; thus

$$e^{-\frac{E_i}{\eta_i}t_0 + c} = -\frac{\sigma_0}{E_T} \tag{8.2.3.11}$$

substituting the above into (8.2.3.10) produces

$$\varepsilon_{Ai} = \frac{\sigma_0}{E_T}\left[1 - e^{-\frac{E_i}{\eta_i}(t-t_0)}\right] \quad i = 1,\ldots,n \tag{8.2.3.12}$$

inserting this expression into equation (8.2.3.1b) associated with the stress equilibrium gives

$$\sigma = \sigma_E + \sum_{i=1}^{n} \sigma_i = E_E \varepsilon_0 + \sum_{i=1}^{n} E_i (\varepsilon_0 - \varepsilon_{Ai}) = E_T \varepsilon_0 - \sum_{i=1}^{n} E_i \varepsilon_{Ai} \qquad (8.2.3.13)$$

and

$$\sigma = \sigma_0 - \sum_{i=1}^{n} E_i \frac{\sigma_0}{E_T} \left[1 - e^{-\frac{E_i}{\eta_i}(t-t_0)} \right] \qquad (8.2.3.14)$$

regrouping terms, one finds

$$\sigma = \sigma_0 \left\{ 1 - \sum_{i=1}^{n} \frac{E_i}{E_T} \left[1 - e^{-\frac{E_i}{\eta_i}(t-t_0)} \right] \right\} = \sigma_0 \frac{E_E}{E_T} \left[1 + \sum_{i=1}^{n} \frac{E_i}{E_E} e^{-\frac{E_i}{\eta_i}(t-t_0)} \right] \qquad (8.2.3.15)$$

the graph that represents this mechanism in figure 8.2.3.3 is similar to that which appeared in figure 8.2.2.5. The stress value varies between the initial value σ_0 and the final value, which is

$$\sigma_\infty = \frac{E_E}{E_T} \sigma_0 \qquad (8.2.3.16)$$

This is a generalization of the result obtained in equation (8.2.2.39).

The expression for the creep function cannot be obtained analytically; thus, the formulation for a material with a model of this kind must be carried out numerically.

Similarly to that defined above, one can define Voigt's generalized standard linear model by joining in series a linear spring and various elements of Voigt, as shown in figure 8.2.3.4.

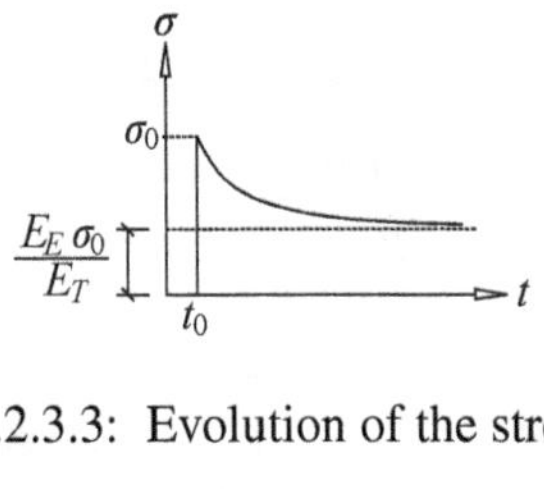

Figure 8.2.3.3: Evolution of the stress value

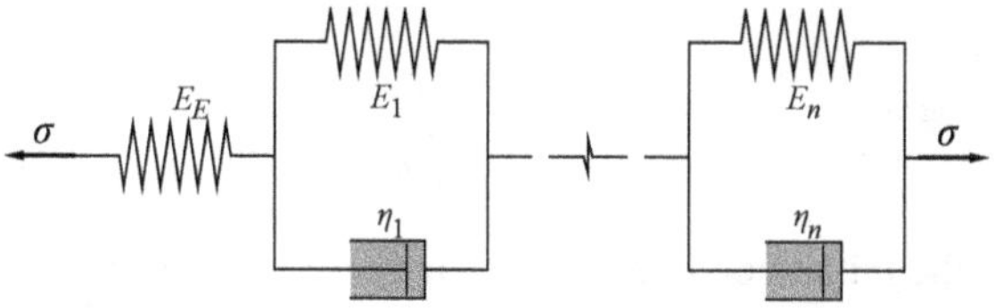

Figure 8.2.3.4: Voigt's generalized linear standard model

According to figure 8.2.3.5, in each element of Voigt's model the following relationships between the stresses and strains can be established

$$\sigma_i = \sigma = \sigma_{Ai} + \sigma_{Mi} \qquad \varepsilon_i = \varepsilon_{Ai} = \varepsilon_{Mi} \tag{8.2.3.17a}$$

$$\sigma_{Mi} = E_i \varepsilon_i \qquad \sigma_{Ai} = \eta_i \dot{\varepsilon}_{Ai} \tag{8.2.3.17b}$$

the total strain is

$$\varepsilon = \frac{\sigma}{E_E} + \sum \varepsilon_i \tag{8.2.3.18}$$

If one wants to study the evolution of the strain that will appear over time in a material when solicited by a constant stress σ_0, one has to study the creep phenomenon. The analysis can start by differentiating with respect to time the strain expression defined in (8.2.3.17a), which gives

$$\dot{\varepsilon}_i = \dot{\varepsilon}_{Ai} = \frac{\sigma_{Ai}}{\eta_i} = \frac{\sigma_0 - \sigma_{Mi}}{\eta_i} = \frac{\sigma_0 - E_i \varepsilon_i}{\eta_i} \tag{8.2.3.19}$$

or

$$\frac{d\varepsilon_i}{\varepsilon_i - \dfrac{\sigma_0}{E_i}} = -\frac{E_i}{\eta_i} dt \tag{8.2.3.20}$$

performing the integration gives

$$\ln\left(\varepsilon_i - \frac{\sigma_0}{E_i}\right) = -\frac{E_i}{\eta_i} t + c \tag{8.2.3.21}$$

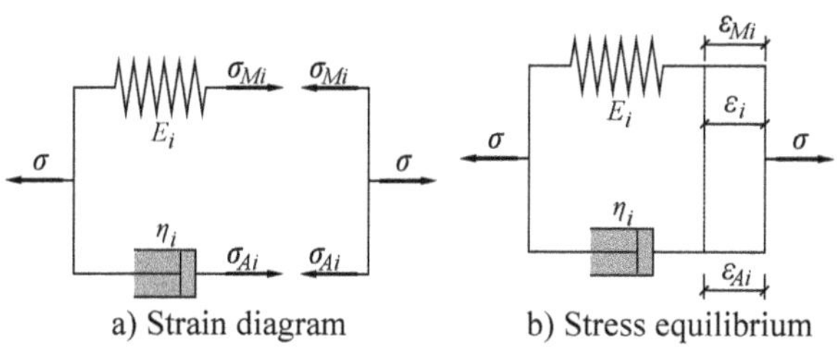

Figure 8.2.3.5: Stresses and strains in a generic element with Voigt's model

or

$$\varepsilon_i - \frac{\sigma_0}{E_{Mi}} = e^{-\frac{E_i}{\eta_i}t + c} \tag{8.2.3.22}$$

Imposing the initial conditions $\varepsilon_i = 0$ and $t = t_0$ results in

$$e^c = -\frac{\sigma_0}{E_i} e^{\frac{E_i}{\eta_i}t_0} \tag{8.2.3.23}$$

finally, one obtains

$$\varepsilon_i = \frac{\sigma_0}{E_i}\left[1 - e^{-\frac{E_i}{\eta_i}(t-t_0)}\right] \tag{8.2.3.24}$$

Therefore, the expression for the total strain is

$$\varepsilon = \sigma_0 \left\{ \frac{1}{E_E} + \sum \frac{1}{E_i}\left[1 - e^{-\frac{E_i}{\eta_i}(t-t_0)}\right]\right\} \tag{8.2.3.25}$$

This expression for $t = t_0$ and for infinite time gives the following values

$$\varepsilon_0 = \frac{\sigma_0}{E_E} \tag{8.2.3.26}$$

$$\varepsilon_\infty = \sigma_0 \left(\frac{1}{E_E} + \sum \frac{1}{E_i} \right) \tag{8.2.3.27}$$

This last expression corresponds to the stiffness of the complete set of elastic springs arranged in series. Figure 8.2.3.6 shows the evolution of the strain over time.

In this model the relaxation function cannot be obtained analytically; thus, it must be computed numerically.

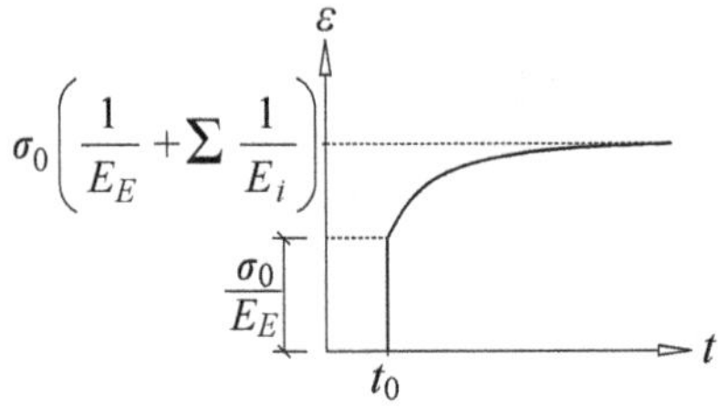

Figure 8.2.3.6: Time evolution of the strain

EXERCISES

8.1. A reinforced concrete beam has a vertical load at its end with the value $P = 40$ kN. The characteristic stress of the concrete is $f_{ck} = 60$ MPa. By monitoring the structure, it has been measured that the vertical displacement at two different times has the following values:

$$t = 1000 \text{ h} \quad w = 7 \text{ mm}$$
$$t = 10000 \text{ h} \quad w = 14 \text{ mm}$$

Find:

1. The values of the parameters of the SLS model that would define the creep function of this beam.
2. The vertical displacement for a time $t = 15000$ h supposing that at 10000 h the concentrated load increases up to a value of 60 kN.

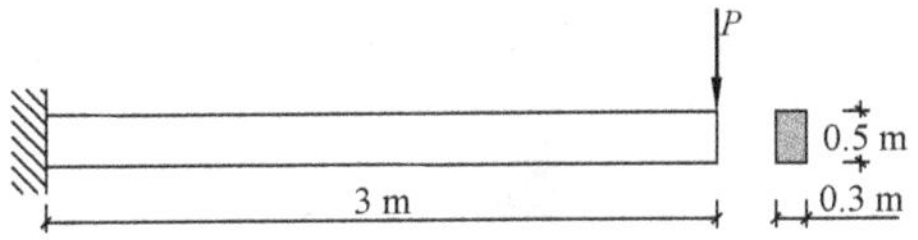

8.2. A cable of a cable-stayed bridge has been prestressed to an initial value of 800 MPa. By monitoring the structure, the following stress values have been found at times t_1 and t_2:

$$t_1 = 1000 \text{ h} \quad \sigma_1 = 790 \text{ MPa}$$
$$t_2 = 5000 \text{ h} \quad \sigma_2 = 760 \text{ MPa}$$

Identify:

1. The coefficients that define the stress expression in the cable over time, assuming that the material behaves according to the viscoelastic SLS model.
2. The stress value to be initially applied to the cable, taking into account the losses due to relaxation.

8.3. The following figure displays a model representing the viscoelastic behaviour of a material.
Find:

1. The constitutive equations of this model.
2. The creep and relaxation functions.

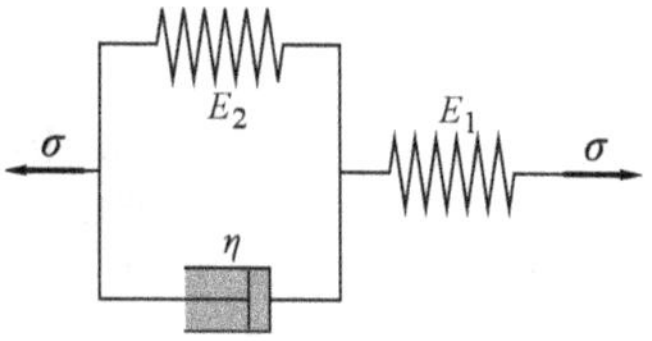

8.4. The figure below describes the concrete beam of a bridge deck with a length of 30 m and supported at both ends. It is loaded with a uniform load, which consists of its weight and the equivalent part of the deck with a total value $p = 40$ kN/m. Also, at the centre of the span, it presents a concentrated load $F = 200$ kN.

By monitoring the structure, the values of the vertical displacement at the centre of the span have been obtained as follows:

$$\text{at } t = 1000 \text{ h} \quad w = 10 \text{ cm}$$
$$\text{at } t = 5000 \text{ h} \quad w = 12 \text{ cm}$$

Assuming that concrete is a viscoelastic material that follows the SLS model, find:

1. The material's creep function.
2. The value of the vertical displacement at $t = 7000$ h at the moment of removing the load F.

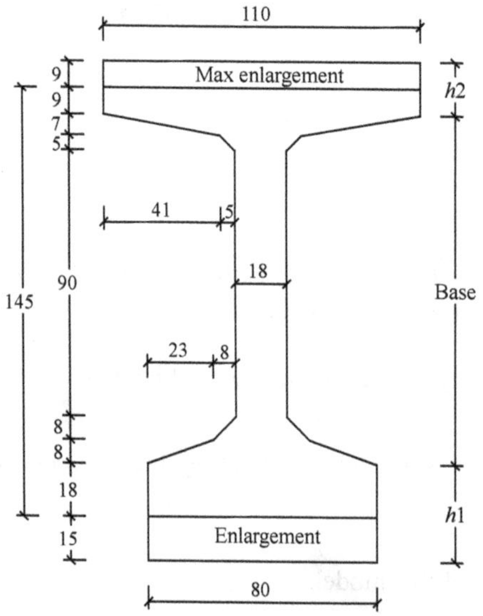

$E = 30000$ MPa
$I = 0.3$ m^4

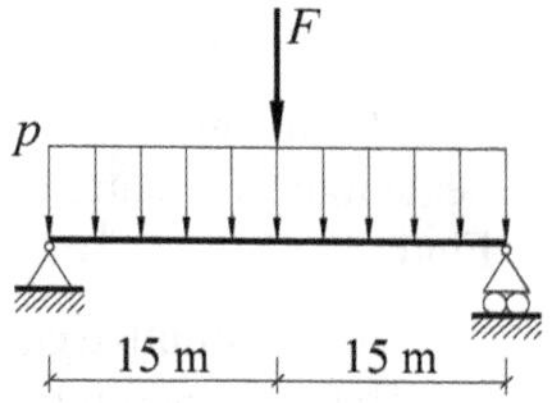

8.5. The following figure presents two models that simulate a material's viscoelastic behaviour, which are usually known as anti-Zener. One of them consists of a Voigt's model connected in series with a linear damper, and the other consists of a Maxwell's model connected in parallel with a linear damper.

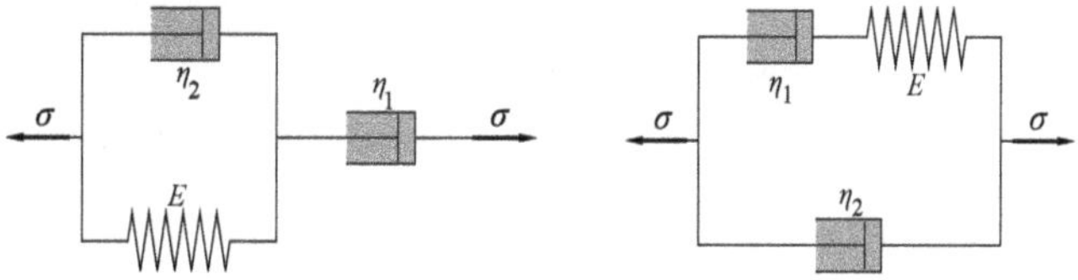

For each of these models, find:

1. The strain expression over time for a constant stress σ_0 and construct its graphical representation.
2. The stress expression in each of the models' branches and construct its graphical representation.

CHAPTER 9

LINEAR ELASTO-VISCOPLASTICITY

9.1 INTRODUCTION

In the previous chapters, different behaviours of materials have been described, and one has been able to see the component of the elastic strain, common to deformable solid bodies, and the component of the viscosity and the plasticity that produce material flow and which are typical of fluids.

These last two behaviours are produced when the material has mechanical properties that vary with time, as happens with viscosity, or when the material reaches a limit, as is the case with plasticity.

There are materials in which these three behaviours are present, that is to say, when the load increases with time, elastic strains initially occur, but once the upper bound of the stress is reached, a continuous strain field appears that combines viscosity and plasticity. It is therefore elasto-viscoplastic behaviour. Fresh concrete and clay soils are examples of these materials.

Different numerical models that attempt to represent the mechanical behaviour of these materials have been developed. This chapter presents the best known.

9.2 MODEL WITH TWO PARAMETERS

This model is composed of the two elements that appear in figure 9.2.1. They are an elastic spring with stiffness E, a linear viscous damper defined by its viscosity η and a fuse that is able to work up to the stress value σ_e.

In this model, the elastic spring and the viscous damper are separated by an element that will hereinafter be referred to as a fuse, which supports the stress,

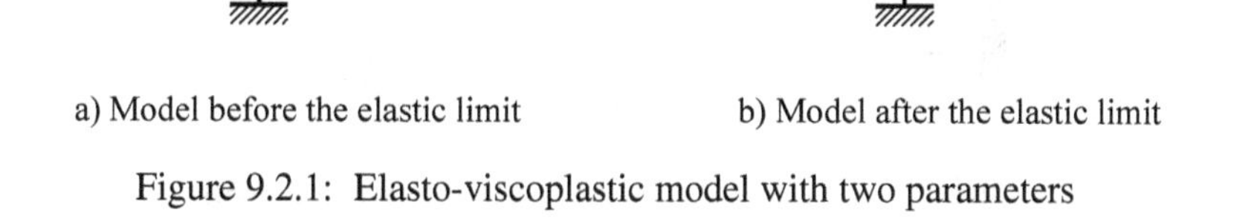

a) Model before the elastic limit b) Model after the elastic limit

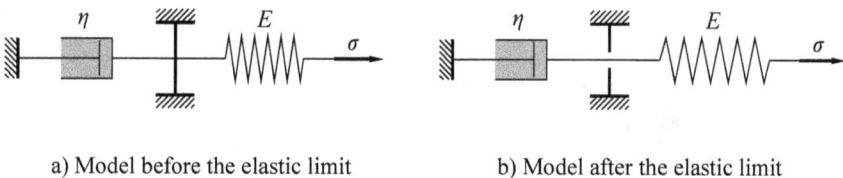

Figure 9.2.1: Elasto-viscoplastic model with two parameters

so that the damper does not receive any. On the other hand, after the fuse is exhausted, due to reaching the material's stress limit, all of the acting stress goes to the damper. Both situations are illustrated in figure 9.2.2.

For the stress reflected in figure 9.2.2, the existing strain is

$$\varepsilon = \frac{\sigma}{E} \tag{9.2.1}$$

where $\sigma < \sigma_e$.

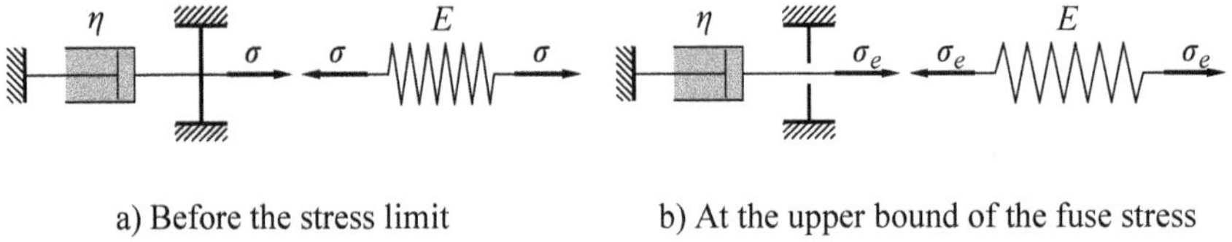
η
E
σ
σ
σ
η
E
σ_e
σ_e
σ_e
a) Before the stress limit
b) At the upper bound of the fuse stress

Figure 9.2.2: Stress equilibrium from the model with two parameters

Once σ_e is reached, the total strain appears as represented in figure 9.2.3.

If one calls ε_A the strain of the viscous damper and ε_e the strain produced by stress σ_e in the elastic spring, which will be constant, one has

$$\varepsilon = \varepsilon_A + \varepsilon_e \tag{9.2.2}$$

calculating the derivative with respect to time gives

$$\dot{\varepsilon} = \dot{\varepsilon}_A = \frac{\sigma_e}{\eta} \tag{9.2.3}$$

and therefore

$$d\varepsilon_A = \frac{\sigma_e}{\eta} dt \tag{9.2.4}$$

integrating (9.2.4) results in

$$\varepsilon_A = \frac{\sigma_e}{\eta} t + c \tag{9.2.5}$$

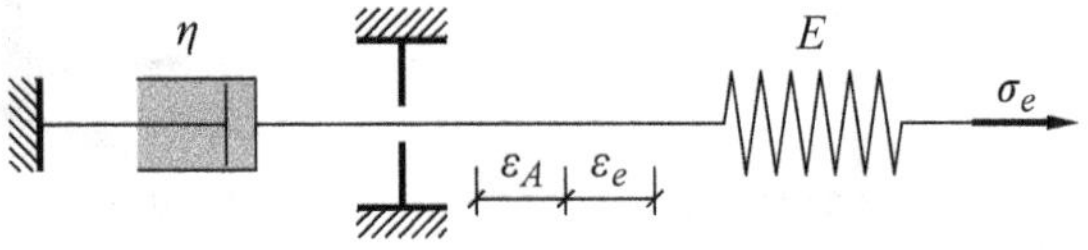

Figure 9.2.3: Strain after the stress limit in the two-parameter model

for $t = t_0$, $\varepsilon_A = 0$, then

$$c = -\frac{\sigma_e}{\eta} t_0 \tag{9.2.6}$$

substituting (9.2.5) and (9.2.2) into the above equation gives

$$\varepsilon = \frac{\sigma_e}{\eta}(t - t_0) + \frac{\sigma_e}{E} \tag{9.2.7}$$

Thus, from (9.2.1) and (9.2.7), one obtains the stress–strain graphic shown in figure 9.2.4.

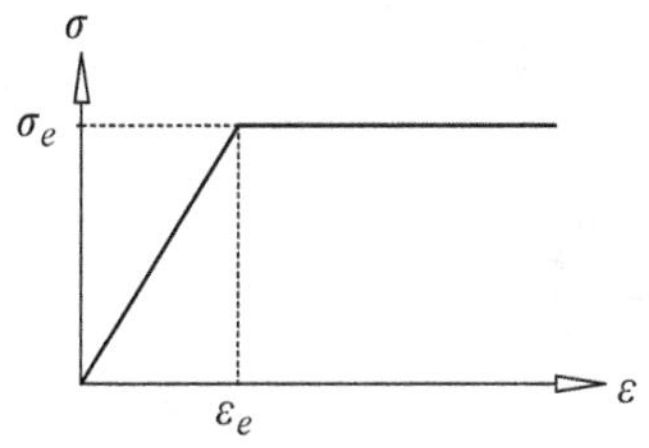

Figure 9.2.4: Graphic of (σ, ε) in the two-parameter model

9.3 MODELS WITH THREE PARAMETERS. BINGHAM MODELS

The model from the previous section does not allow stress increases from σ_e onwards. However, there are materials in which strain hardening occurs due to the elastic strain; thus, its strength capabilities are improved. To represent such behaviour it is necessary to modify the former model. A well-known alternative was established by Bingham, who formulated a model consisting of an elastic spring, a viscous damper and a friction element that works up to strain ε_e. This model is represented in figure 9.3.1.

The model's behaviour depends on the stress value σ. The damper does not support any stress until the strain reaches the value ε_e and then

$$\varepsilon = \frac{\sigma}{E} \tag{9.3.1}$$

when $\sigma < \sigma_e$.

At σ_e the friction element is exhausted, and afterwards the viscous damper supports the stress increase that exceeds σ_e. The expression of the total strain is

$$\varepsilon = \varepsilon_E + \varepsilon_A = \varepsilon_E + \varepsilon_p \tag{9.3.2}$$

with

$$\varepsilon_E = \frac{\sigma}{E} \tag{9.3.3}$$

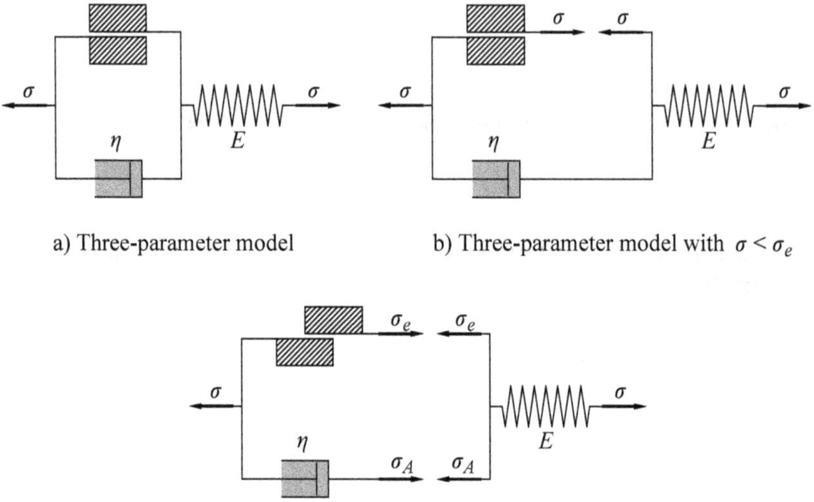

a) Three-parameter model

b) Three-parameter model with $\sigma < \sigma_e$

c) Three-parameter model with $\sigma \geq \sigma_e$

Figure 9.3.1: Bingham's three-parameter model

Figures 9.3.2 and 9.3.3 reflect both cases.

To obtain the creep function, in other words, to obtain the strain produced by a stress with constant value $\sigma > \sigma_e$, it is convenient to start from equation (9.3.2), taking the derivatives with respect to time and recalling that the strain ε_E is constant. By doing that, the result is

$$\dot{\varepsilon} = \dot{\varepsilon}_A = \frac{\sigma - \sigma_e}{\eta} \tag{9.3.4}$$

or

$$d\varepsilon_A = \left(\frac{\sigma - \sigma_e}{\eta}\right) dt \tag{9.3.5}$$

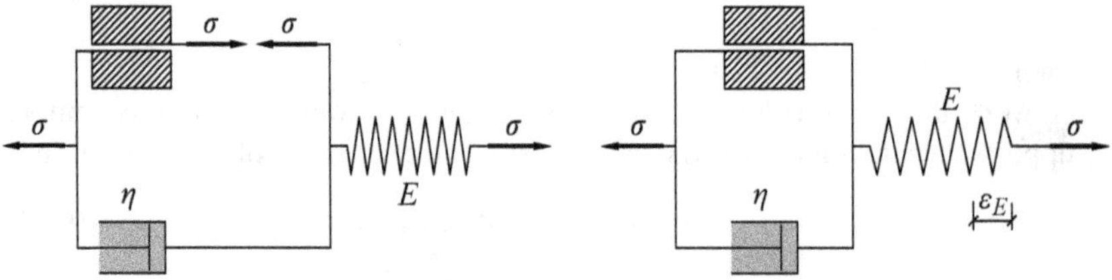

Figure 9.3.2: Three-parameter model for $\sigma < \sigma_e$

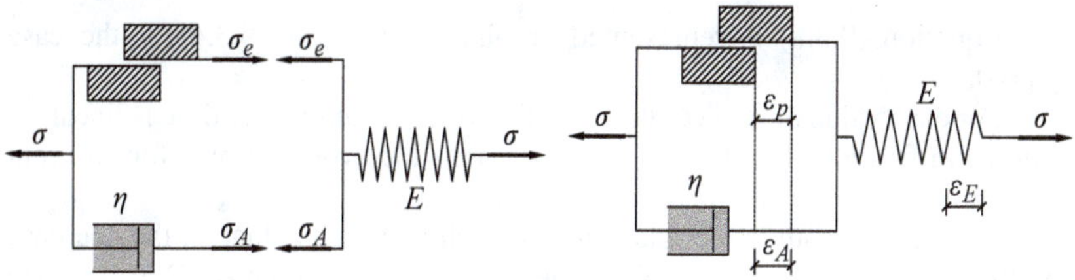

Figure 9.3.3: Three-parameter model for $\sigma \geq \sigma_e$

integrating (9.3.5) gives

$$\varepsilon_A = \frac{\sigma - \sigma_e}{\eta} t + c \tag{9.3.6}$$

at $t = t_0$, one has $\varepsilon_A = 0$. Then

$$c = -\frac{\sigma - \sigma_e}{\eta} t_0 \tag{9.3.7}$$

and

$$\varepsilon = \varepsilon_E + \left(\frac{\sigma - \sigma_e}{\eta} \right) (t - t_0) = \frac{\sigma}{E} + \left(\frac{\sigma - \sigma_e}{\eta} \right) (t - t_0) \tag{9.3.8}$$

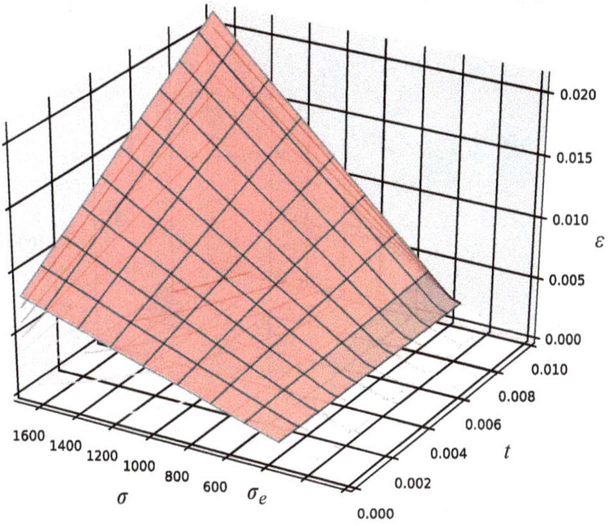

Figure 9.3.4: Graphical representation of $\varepsilon = \varepsilon(\sigma, t)$

Equation (9.3.8) is represented graphically in figure 9.3.4 for the case $t_0 = 0$.

For each value of $\sigma \geq \sigma_e$, the evolution of the strain ε over time is linear, as shown in figure 9.3.5a. Figure 9.3.5b presents the evolution of ε for different values of σ in the example of figure 9.3.4.

One can also study the relaxation of such materials – that is, the decrease in the stress $\sigma_0 \geq \sigma_e$ that initially generates a constant strain ε_0. One can once again recall that

$$\varepsilon = \varepsilon_E + \varepsilon_A \tag{9.3.9}$$

carrying out the derivatives with respect to time

$$\dot{\varepsilon} = \dot{\varepsilon}_E + \dot{\varepsilon}_A = \frac{\dot{\sigma}}{E} + \frac{\sigma - \sigma_e}{\eta} = 0 \tag{9.3.10}$$

and rearranging terms

$$\frac{\dot{\sigma}}{\sigma - \sigma_e} = -\frac{E}{\eta} \tag{9.3.11}$$

This expression can be written as

$$\frac{d\sigma}{\sigma - \sigma_e} = -\frac{E}{\eta}dt \tag{9.3.12}$$

performing the integration results in

$$\ln(\sigma - \sigma_e) = -\frac{E}{\eta}t + c \tag{9.3.13}$$

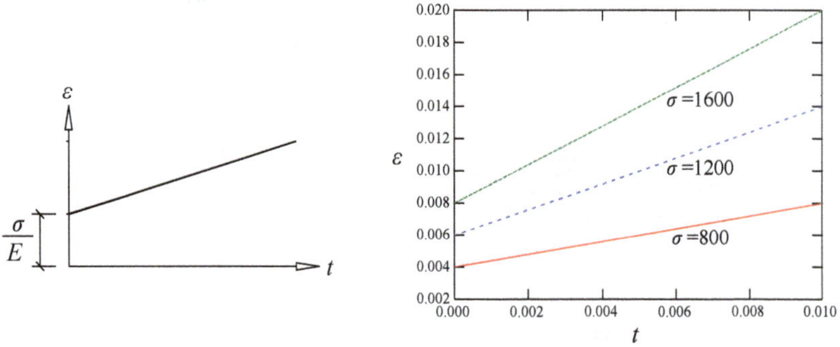

a) Evolution of the strain b) Evolution of ε for different σ

Figure 9.3.5: Time history of the strain

or

$$\sigma - \sigma_e = e^{-\frac{E}{\eta}t + c} \qquad (9.3.14)$$

for $t = t_0$, $\sigma = \sigma_0$; thus

$$(\sigma_0 - \sigma_e)\, e^{\frac{E}{\eta}t_0} = e^c \qquad (9.3.15)$$

and therefore one finally obtains

$$\sigma = \sigma_e + (\sigma_0 - \sigma_e)\, e^{-\frac{E}{\eta}(t-t_0)} = \sigma_0 - (\sigma_0 - \sigma_e)\left[1 - e^{-\frac{E}{\eta}(t-t_0)}\right] \qquad (9.3.16)$$

Figure 9.3.6 presents the evolution in time of σ resulting from equation (9.3.16).

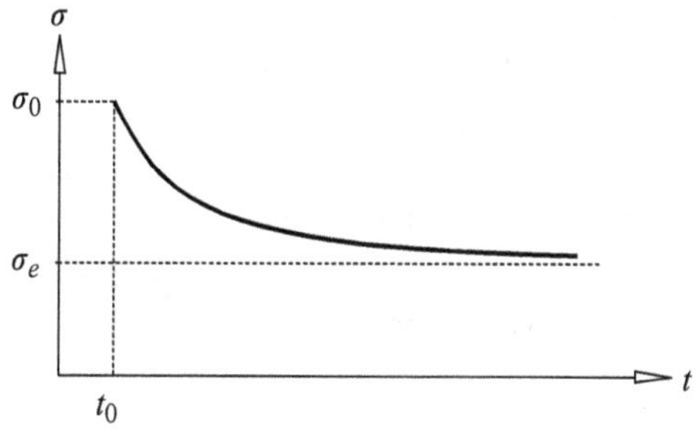

Figure 9.3.6: Evolution of stress under constant strain

9.4 BINGHAM'S FOUR-PARAMETER MODEL

This model incorporates a second elastic and linear spring that preserves part of the elastic behaviour after the exhaustion of the friction element. The model is described in figure 9.4.1.

The strains that occur in the model at different values of σ are shown in figure 9.4.2.

The procedure to obtain the creep function in this model is similar to that in the previous section. If the stress exceeds the value of σ_e, the strain will be

$$\varepsilon = \varepsilon_1 + \varepsilon_p = \varepsilon_1 + \varepsilon_A = \varepsilon_1 + \varepsilon_2 \qquad (9.4.1)$$

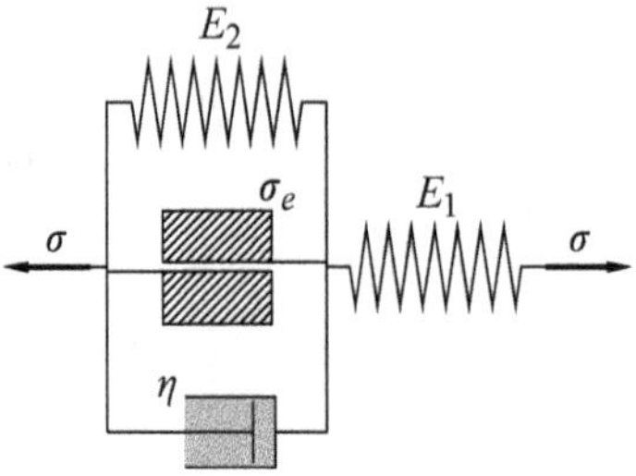

a) Four-parameter model

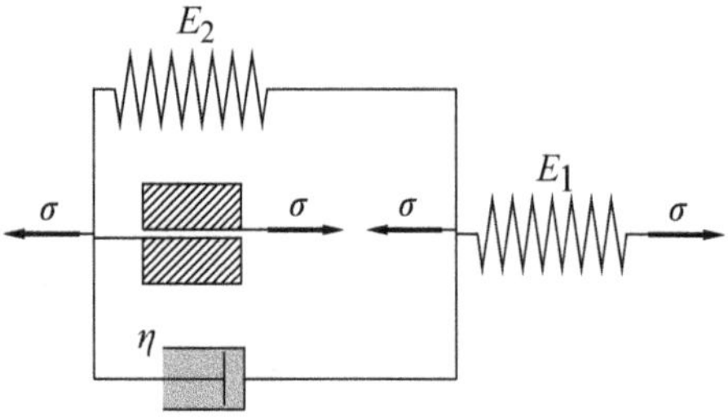

b) Stress equilibrium for $\sigma < \sigma_e$

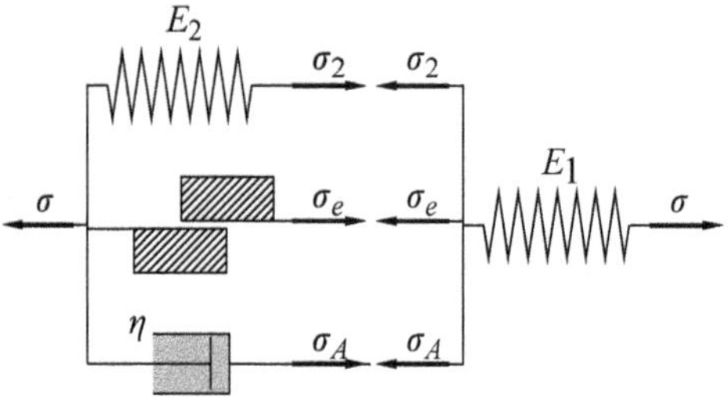

c) Stress equilibrium for $\sigma \geq \sigma_e$

Figure 9.4.1: Bingham's four-parameter model

and the stress equilibrium will become

$$\sigma = \sigma_e + \sigma_2 + \sigma_A \tag{9.4.2}$$

If the load is constant, ε_1 is also constant, and so differentiating (9.4.1) with respect to time gives

$$\dot{\varepsilon} = \dot{\varepsilon}_A = \frac{\sigma - \sigma_e - \sigma_2}{\eta} = \frac{\sigma - \sigma_e - E_2 \varepsilon_A}{\eta} \tag{9.4.3}$$

rearranging terms produces

$$\frac{\dot{\varepsilon}_A}{\varepsilon_A + \dfrac{\sigma_e - \sigma}{E_2}} = -\frac{E_2}{\eta} \tag{9.4.4}$$

or

$$\frac{d\varepsilon_A}{\varepsilon_A - \dfrac{\sigma - \sigma_e}{E_2}} = -\frac{E_2}{\eta} dt \tag{9.4.5}$$

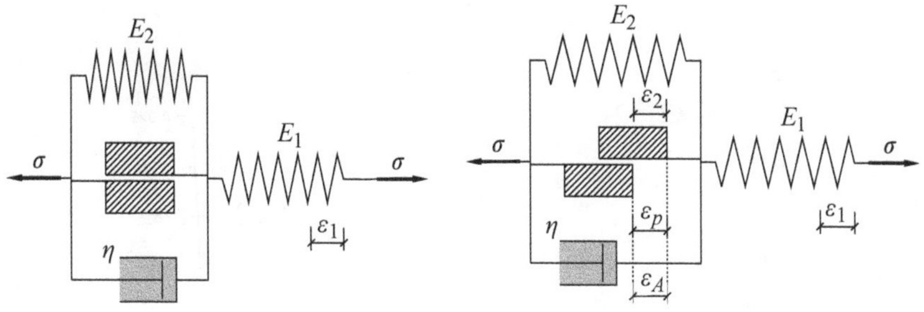

a) Strain for $\sigma < \sigma_e$ b) Strain for $\sigma \geq \sigma_e$

Figure 9.4.2: Strain cases in the four-parameter model

performing the integration gives

$$\ln\left(\varepsilon_A - \frac{\sigma - \sigma_e}{E_2}\right) = -\frac{E_2}{\eta}t + c \tag{9.4.6}$$

which can also be written as

$$\varepsilon_A - \frac{\sigma - \sigma_e}{E_2} = e^{-\frac{E_2}{\eta}t + c} \tag{9.4.7}$$

for $t = t_0$, $\varepsilon_A = 0$; thus

$$e^c = -\left(\frac{\sigma - \sigma_e}{E_2}\right) e^{\frac{E_2}{\eta}t_0} \tag{9.4.8}$$

and therefore

$$\varepsilon_A = \frac{\sigma - \sigma_e}{E_2} - \frac{\sigma - \sigma_e}{E_2} e^{-\frac{E_2}{\eta}(t-t_0)} \tag{9.4.9}$$

substituting in (9.4.1) finally gives

$$\varepsilon = \frac{\sigma}{E_1} + \frac{\sigma - \sigma_e}{E_2}\left[1 - e^{-\frac{E_2}{\eta}(t-t_0)}\right] \tag{9.4.10}$$

The graphical representation of this expression for $\sigma \geq \sigma_e$ is shown in figure 9.4.3.

For constant stress values, the evolution of the strain is shown in figure 9.4.4a, and the results corresponding to several values of σ in the example of figure 9.4.3 can be seen in figure 9.4.4b.

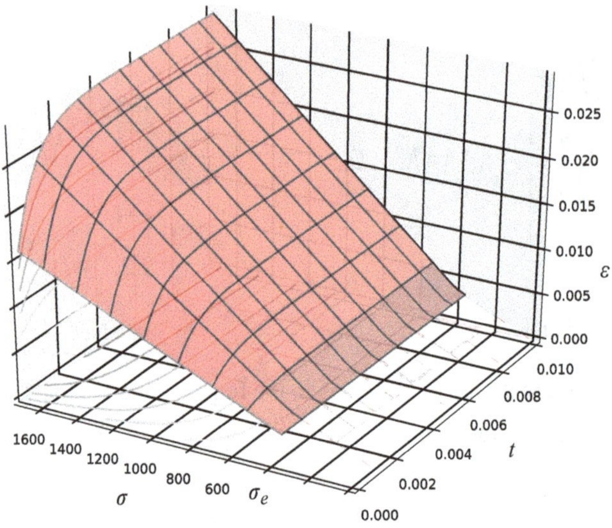

Figure 9.4.3: Graphical representation of $\varepsilon = \varepsilon(\sigma, t)$

To obtain the relaxation function – that is to say, the time history of the stress $\sigma_0 \geq \sigma_e$ that produces a constant strain ε_0 – one starts again from the strain expression, i.e.

$$\varepsilon = \varepsilon_1 + \varepsilon_2 = \frac{\sigma}{E_1} + \frac{\sigma_2}{E_2} \tag{9.4.11}$$

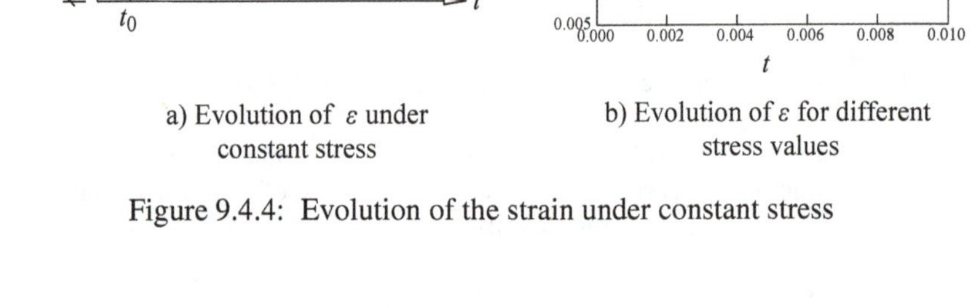

a) Evolution of ε under constant stress

b) Evolution of ε for different stress values

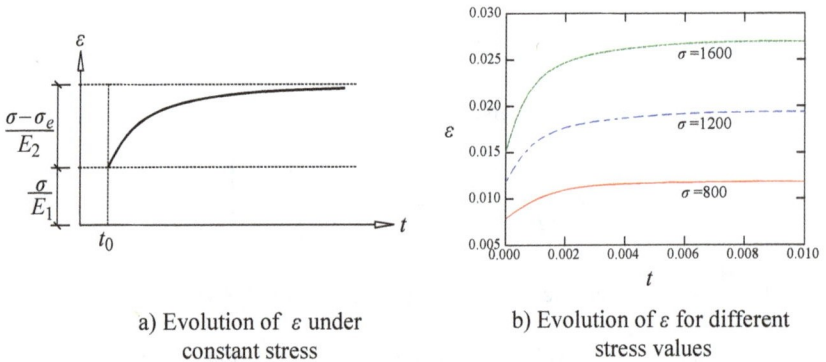

Figure 9.4.4: Evolution of the strain under constant stress

differentiating with respect to time gives

$$\dot{\varepsilon} = \frac{\dot{\sigma}}{E_1} + \frac{\dot{\sigma}_2}{E_2} = 0 \tag{9.4.12}$$

or

$$\frac{\dot{\sigma}}{E_1} = -\frac{\dot{\sigma}_2}{E_2} \tag{9.4.13}$$

integrating (9.4.13) results in

$$\frac{\sigma}{E_1} = -\frac{\sigma_2}{E_2} + c \tag{9.4.14}$$

for $t = t_0$, $\sigma = \sigma_0$ and $\sigma_2 = 0$; thus

$$\frac{\sigma_0}{E_1} = c \tag{9.4.15}$$

substituting in (9.4.14) gives

$$\sigma_2 = \frac{E_2}{E_1}(\sigma_0 - \sigma) \tag{9.4.16}$$

one also knows that

$$\varepsilon = \varepsilon_1 + \varepsilon_A \tag{9.4.17}$$

differentiating this expression with respect to time gives

$$\dot{\varepsilon} = \dot{\varepsilon}_1 + \dot{\varepsilon}_A = \frac{\dot{\sigma}}{E_1} + \frac{\dot{\sigma}_A}{\eta} = \frac{\dot{\sigma}}{E_1} + \frac{\sigma - \sigma_2 - \sigma_e}{\eta} = 0 \tag{9.4.18}$$

substituting for σ_2 in equation (9.4.16) results in

$$\dot{\sigma} = -\frac{1}{\eta}\left[(E_1 + E_2)\,\sigma - E_2\sigma_0 - E_1\sigma_e\right] \tag{9.4.19}$$

and rearranging the terms gives

$$\frac{\dot{\sigma}}{\sigma - \dfrac{E_2\sigma_0 + E_1\sigma_e}{E_1 + E_2}} = -\frac{E_1 + E_2}{\eta} \tag{9.4.20}$$

or

$$\frac{d\sigma}{\sigma - \dfrac{E_2\sigma_0 + E_1\sigma_e}{E_1 + E_2}} = -\frac{E_1 + E_2}{\eta}dt \tag{9.4.21}$$

performing the integration results in

$$\ln\left(\sigma - \frac{E_2\sigma_0 + E_1\sigma_e}{E_1 + E_2}\right) = -\frac{E_1 + E_2}{\eta}t + c \tag{9.4.22}$$

or

$$\sigma - \frac{E_2\sigma_0 + E_1\sigma_e}{E_1 + E_2} = e^{-\frac{E_1 + E_2}{\eta}t + c} \tag{9.4.23}$$

for $t = t_0$, $\sigma = \sigma_0$; thus

$$e^c = \frac{E_1(\sigma_0 - \sigma_e)}{E_1 + E_2} e^{\frac{E_1 + E_2}{\eta}t_0} \tag{9.4.24}$$

substituting the above formula into (9.4.23), one finally obtains

$$\sigma = \sigma_0 - \frac{E_1(\sigma_0 - \sigma_e)}{E_1 + E_2}\left[1 - e^{-\frac{E_1 + E_2}{\eta}(t - t_0)}\right] \tag{9.4.25}$$

The graphical representation of this expression is shown in figure 9.4.5.

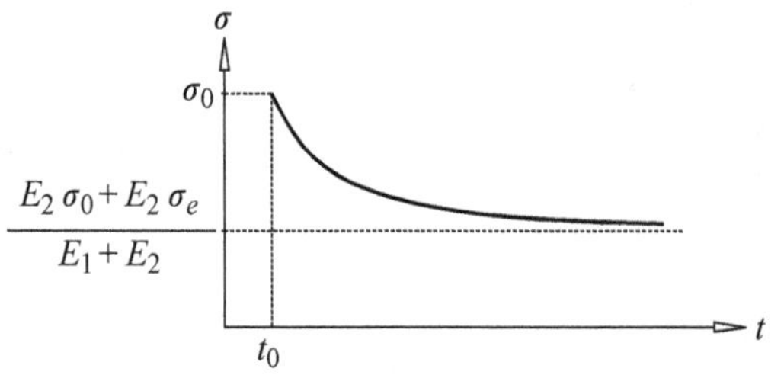

Figure 9.4.5: Time evolution of the stress under constant strain

9.5 NON-LINEAR VISCOSITY MODELS. NORTON'S MODEL

In the previous paragraphs the viscous damper has been considered to have a constant viscosity coefficient η; however, the approach of Norton considers the viscosity coefficient as a function of the damper stress. The general expression is

$$\eta = \frac{\lambda^N}{\sigma_A^{N-1}} \tag{9.5.1}$$

where σ_A is the stress in the damper, λ is a viscosity coefficient with constant value, and N is the exponent of the expression, which must be found through laboratory tests. As a result, the constitutive equation of a non-linear viscous damper is

$$\dot{\varepsilon}_A = \frac{\sigma^N}{\lambda^N} \tag{9.5.2}$$

Substituting this expression into the equations of the preceding paragraphs, one obtains complex formulations that usually require a numerical solution.

EXERCISES

9.1 In the following figure a model is shown, known as the BingMax model for an elasto-viscoplastic material. Find:

1. The model for the material when $E_1 \to \infty$.
2. The model for the material when $E_2 \to \infty$.
3. The model for the material when $\eta \to \infty$.
4. The model for the material when $\sigma_e = 0$.
5. The creep and relaxation functions for the general case in which all the model's parameters have finite values.

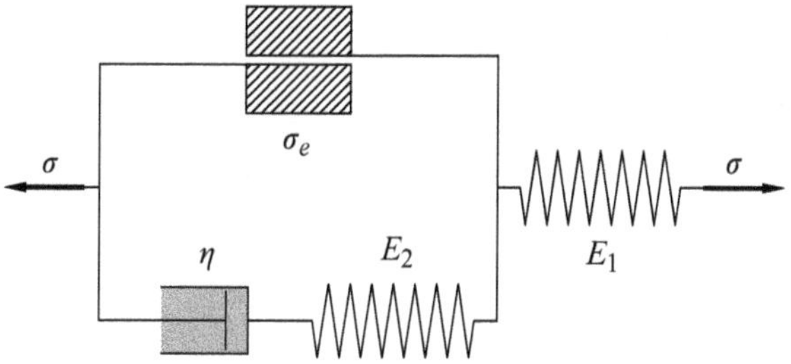

9.2 The model in the following figure is called the Nishihara formulation and it is commonly used to study the consolidation process of clay soils. As can be seen, it consists of a linear elastic spring, a Voigt module and a Bingham module. Find the creep function associated with this elasto-viscoplastic model.

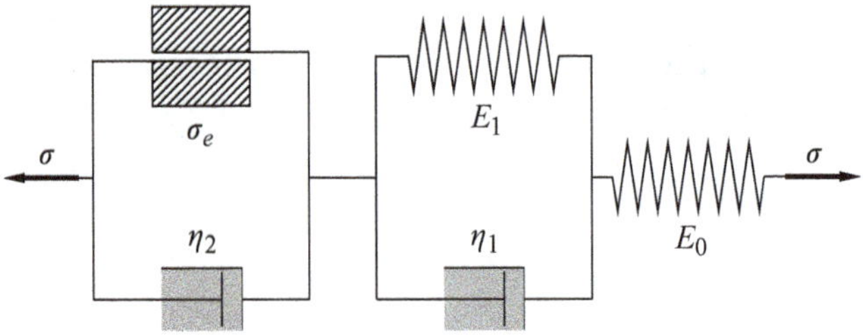

9.3 In the following figure the results for the evolution of the strain over time of three specimens of an elasto-viscoplastic material that have been tested at different stress values are shown.
Find:

1. The mathematical model that is more suitable for this material.
2. The values of the parameters that define the creep function and its expression.

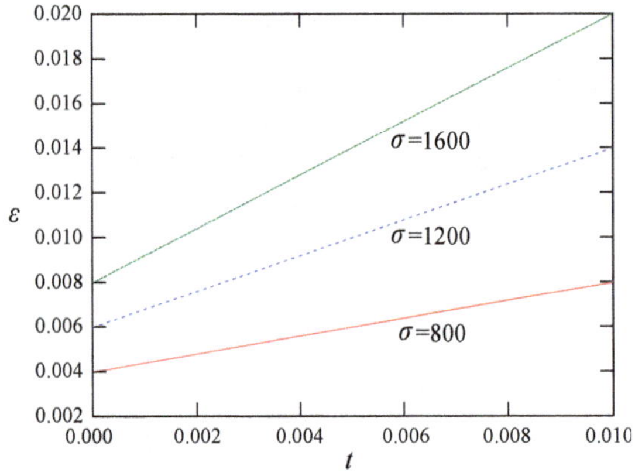

BIBLIOGRAPHY

Díaz del Valle, J. (1989). *Mecánica de los Medios Continuos*. E.T.S. Ingenieros de Caminos, Canales y Puertos, Santander.

Holzapfel, G.A. (2010). *Nonlinear Solid Mechanics: A Continuum Approach for Engineering*. John Wiley, Chichester.

Lai, W.M., Rubin, D., Krempl, E. (2010). *Introduction to Continuum Mechanics*. Elsevier, Burlington, MA/Oxford.

Lubarda, V.A. (2002). *Elastoplasticity Theory*. CRC Press LLC, Washington, D.C.

Ogden, R.W. (1997). *Non-Linear Elastic Deformations*. Dover Publications, New York.

Pastoriza, A., García, L., Núñez, A. (1970). *Elementos de Elasticidad*. E.T.S. Ingenieros de Caminos, Canales y Puertos, Santander.

Samartín, A. (1990). *Curso de Elasticidad*. Bellisco, Madrid.

Shabana, A.A. (2012). *Computational Continuum Mechanics*. Cambridge University Press, New York.

Timoshenko, S., Goodier, J.N. (1970). *Theory of Elasticity*. McGraw-Hill, New York.

www.ingramcontent.com/pod-product-compliance
Lightning Source LLC
Chambersburg PA
CBHW050451190326
41458CB00005B/1234
9781784662714